AF806259

Socio-Affective Computing

Volume 1

Series Editor
Amir Hussain, University of Stirling, Stirling, UK

Co Editor
Erik Cambria, Nanyang Technological University, Singapore

This exciting Book Series aims to publish state-of-the-art research on socially intelligent, affective and multimodal human-machine interaction and systems. It will emphasize the role of affect in social interactions and the humanistic side of affective computing by promoting publications at the cross-roads between engineering and human sciences (including biological, social and cultural aspects of human life). Three broad domains of social and affective computing will be covered by the book series: (1) social computing, (2) affective computing, and (3) interplay of the first two domains (for example, augmenting social interaction through affective computing). Examples of the first domain will include but not limited to: all types of social interactions that contribute to the meaning, interest and richness of our daily life, for example, information produced by a group of people used to provide or enhance the functioning of a system. Examples of the second domain will include, but not limited to: computational and psychological models of emotions, bodily manifestations of affect (facial expressions, posture, behavior, physiology), and affective interfaces and applications (dialogue systems, games, learning etc.). This series will publish works of the highest quality that advance the understanding and practical application of social and affective computing techniques. Research monographs, introductory and advanced level textbooks, volume editions and proceedings will be considered.

More information about this series at http://www.springer.com/series/13199

Erik Cambria • Amir Hussain

Sentic Computing

A Common-Sense-Based Framework for Concept-Level Sentiment Analysis

Source is: www.sentic.net
Copyrights Erik Cambria, Amir Hussain

Springer

MÄNGELEXEMPLAR

Erik Cambria
School of Computer Engineering
Nanyang Technological University
Singapore, Singapore

Amir Hussain
Computing Science and Mathematics
University of Stirling
Stirling, UK

Socio-Affective Computing
ISBN 978-3-319-23653-7 ISBN 978-3-319-23654-4 (eBook)
DOI 10.1007/978-3-319-23654-4

Library of Congress Control Number: 2015950064

Springer Cham Heidelberg New York Dordrecht London
© Springer International Publishing Switzerland 2015
This work is subject to copyright. All rights are reserved by the Publisher, whether the whole or part of the material is concerned, specifically the rights of translation, reprinting, reuse of illustrations, recitation, broadcasting, reproduction on microfilms or in any other physical way, and transmission or information storage and retrieval, electronic adaptation, computer software, or by similar or dissimilar methodology now known or hereafter developed.
The use of general descriptive names, registered names, trademarks, service marks, etc. in this publication does not imply, even in the absence of a specific statement, that such names are exempt from the relevant protective laws and regulations and therefore free for general use.
The publisher, the authors and the editors are safe to assume that the advice and information in this book are believed to be true and accurate at the date of publication. Neither the publisher nor the authors or the editors give a warranty, express or implied, with respect to the material contained herein or for any errors or omissions that may have been made.

Printed on acid-free paper

Springer International Publishing AG Switzerland is part of Springer Science+Business Media (www.springer.com)

In memory of Dustin,

A man with a great mind and a big heart.

Foreword

It was a particular joy to me having been asked to write a few words for this second book on sentic computing - the first book published in 2012, gave me immense inspiration which has gripped me ever since. This also makes it a relatively easy bet that it will continue its way to a standard reference that will help change the way we approach sentiment, emotion, and affect in natural language processing and beyond.

While approaches to integrate emotional aspects in natural language understanding date back to the early 1980s such as in Dyer's work on *In-Depth Understanding*, at the very turn of the last millennium, there was still very limited literature in this direction. It was about 3 years after Picard's 1997 field-defining book on *Affective Computing* and one more after the first paper on *Recognizing Emotion in Speech* by Dellaert, Polzin, and Waibel and a similar one by Cowie and Douglas-Cowie that followed ground-laying work, including by Scherer and colleagues on vocal expression of emotion and earlier work on synthesizing emotion in speech when a global industrial player placed an order for a study whether we can enable computers to recognize users' emotional factors in order to make human-computer dialogues more natural.

After first attempts to grasp emotion from facial expression, our team realized that computer vision was not truly ready back then for "in the wild" processing. Thus, the thought came to mind to train our one-pass top-down natural language understanding engine to recognize emotion from speech instead. In doing so, I was left with two options: train the statistical language model or the acoustic model to recognize basic emotions rather than understand spoken content. I decided to do both and, alas, it worked – at least to some degree. However, when I presented this new ability, the usual audience response was, mainly along the lines of "*Interesting, but what is the application?*" Since then, a major change of mind has taken place: it is by and large agreed that taking into account emotions is key for natural language processing and understanding, especially for tasks such as sentiment analysis.

As a consequence, these days, several hundred papers dealing with the topic appear annually, and one finds several thousand citations each year in this field which is still gaining momentum and expected to be nothing less than a game-changing factor in addressing future computing challenges, such as when mining opinion, retrieving information, or interacting with technical systems. Hardly surprisingly, the commercial interest is ever rising, and first products-have already found their way into broad public awareness.

The lion's share of today's work aimed at dealing with analyzing emotion and sentiment in spoken and written language, is based on statistical word co-occurrences. The principle is described in Joachim's 1996 work on text categorization representing a document as a "bag-of-words" in a vector space. Different normalizations are named, and sequences of *n* words or characters ('*n*-grams') have since been applied successfully in similar fashion. With the advent of "big data," recent approaches, such as by Google, translate the single words into (their individual) vectors by (some form of) soft clustering. This reflects each word's relation to the other words in the vocabulary as added information. However, such approaches have reached a certain glass ceiling over the years as they are very limited in taking inspiration from how the human brain processes both emotions (by exploiting an emotion model) and meaning (by working at the semantic/concept level rather than at the syntactic/word level) to perform natural language processing tasks such as information extraction and sentiment analysis.

This is what *Sentic Computing* is all about. Targeting the higher hanging fruits by not missing the importance of aiming to emulate the brain's described processing of emotions *and* meaning, it provides a knowledge-based approach to concept-level sentiment analysis that is well rooted in a multi-disciplinary view. The consideration of a text as a bag-of-words is accordingly substituted by representing it as a "bag-of-concepts." This embeds linguistics in an elegant form beyond mere statistics, enriching the representation of text by the dependency relation between clauses. The book guides its readers from the student-level onwards in an ingenious fashion, from an introduction and background knowledge (not only on sentiment analysis and opinion mining but also common sense) to the core piece – SenticNet (introducing the acquisition and representation of knowledge as well as reasoning) to *concept-level* sentiment analysis. It then exemplifies these ideas by three excellently picked applications in the domains of the social web, human-computer interaction, and e-health systems before concluding remarks. Thus, besides providing the essential comprehension of the basics of the field in a smooth and very enjoyable way, it not only manages to take the reader to the next level but introduces genuine novelty of utmost valuable inspiration to any expert in the field. In fact, it makes a major contribution to the next generation of emotionally intelligent computer systems. Do not be surprised catching yourself reasoning about sentiment and opinions in a whole new way even in your "non-tech" life. It remains to say that I am truly looking forward to the volumes to follow this one that kicks off the series on *Socio-Affective Computing* edited by the authors – and sets the bar utmost high for all aspiring readers.

Imperial College, London, UK
July 2015

Björn W. Schuller
President, Association for the Advancement
of Affective Computing (AAAC)
Editor-in-Chief: IEEE Transactions
on Affective Computing

Preface

The opportunity to capture the opinions of the general public has raised growing interest both within the scientific community, leading to many exciting open challenges, and in the business world due to the remarkable range of benefits envisaged, including from marketing, business intelligence and financial prediction. Mining opinions and sentiments from natural language, however, is an extremely difficult task as it involves a deep understanding of most of the explicit and implicit, regular and irregular, syntactical and semantic rules appropriate of a language. Existing approaches to sentiment analysis mainly rely on parts of text in which opinions are explicitly expressed such as polarity terms, affect words, and their co-occurrence frequencies. However, opinions and sentiments are often conveyed implicitly through latent semantics, which make purely syntactical approaches ineffective.

Concept-level approaches, instead, use Web ontologies or semantic networks to accomplish semantic text analysis. This helps the system grasp the conceptual and affective information associated with natural language opinions. By relying on large semantic knowledge bases, such approaches step away from blindly using keywords and word co-occurrence counts and instead rely on the implicit meaning/features associated with natural language concepts. Superior to purely syntactical techniques, concept-based approaches can detect subtly expressed sentiments. Concept-based approaches, in fact, can analyze multi-word expressions that do not explicitly convey emotion, but are related to concepts that do so.

Sentic computing is a pioneering multi-disciplinary approach to natural language processing and understanding at the crossroads between affective computing, information extraction, and common-sense reasoning, and exploits both computer and human sciences to better interpret and process social information on the Web. In sentic computing, whose term derives from the Latin "sentire" (root of words such as sentiment and sentience) and "sensus" (as in common sense), the analysis of natural language is based on common-sense reasoning tools, which enable the analysis of text not only at the document, page, or paragraph level but also at the sentence, clause, and concept level.

This book, a sequel of the first edition published in 2012 as Volume one of SpringerBriefs in Cognitive Computation, focuses on explaining the three key shifts proposed by sentic computing, namely:

1. Sentic computing's *shift from mono- to multi-disciplinarity* – evidenced by the concomitant use of AI and Semantic Web techniques, for knowledge representation and inference; mathematics, for carrying out tasks such as graph mining and multi-dimensionality reduction; linguistics, for discourse analysis and pragmatics; psychology, for cognitive and affective modeling; sociology, for understanding social network dynamics and social influence; and finally ethics, for understanding related issues about the nature of mind and the creation of emotional machines.
2. Sentic computing's *shift from syntax to semantics* – enabled by the adoption of the bag-of-concepts model instead of simply counting word co-occurrence frequencies in text. Working at concept level entails preserving the meaning carried by multi-word expressions such as `cloud_computing`, which represent "semantic atoms" that should never be broken down into single words. In the bag-of-words model, for example, the concept `cloud_computing` would be split into `computing` and `cloud`, which may wrongly activate concepts related to the weather and, hence, compromise categorization accuracy.
3. Sentic computing's *shift from statistics to linguistics* – implemented by allowing sentiments to flow from concept to concept based on the dependency relation between clauses. The sentence "iPhone6 is expensive but nice", for example, is equal to "iPhone6 is nice but expensive" from a bag-of-words perspective. However, the two sentences bear opposite polarity: the former is positive as the user seems to be willing to make the effort to buy the product despite its high price, while the latter is negative as the user complains about the price of iPhone6 although he/she likes it.

Singapore — Erik Cambria
Stirling, UK — Amir Hussain
Sep 2015

Contents

List of Figures

List of Tables

Acronyms

3NF	Third Normal Form
AI	Artificial Intelligence
AKAPU	Average Knowledge Acquired per User
ANN	Artificial Neural Network
API	Application Programming Interface
BACK	Benchmark for Affective Common-Sense Knowledge
BCNF	Boyce-Codd Normal Form
CF-IOF	Concept Frequency-Inverse Opinion Frequency
CMC	Computer-Mediated Communication
CQ	Cultural Quotient
DAG	Directed Acyclic Graph
DAU	Daily Active User
DL	Description Logic
EFACS	Emotional Facial Action Coding System
ELM	Extreme Learning Machine
EQ	Emotional Quotient
FACS	Facial Action Coding System
FMRI	Functional Magnetic Resonance Imaging
FOAF	Friend of a Friend
FOL	First-Order Logic
GECKA	Game Engine for Common-Sense Knowledge Acquisition
GWAP	Game with a Purpose
HCI	Human-Computer Interaction
HEO	Human Emotion Ontology
HMM	Hidden Markov Model
HRQoL	Health-Related Quality of Life
HTML	Hypertext Markup Language
HVS	Human Visual System
IM	Instant Messaging
IQ	Intelligence Quotient
IT	Information Technology

IUI	Intelligent User Interface
JL	Johnson-Lindenstrauss
KNN	K-Nearest Neighbors
KR	Knowledge Representation
JSON	JavaScript Object Notation
LSA	Latent Semantic Analysis
MAU	Monthly Active User
MFCC	Mel Frequency Cepstral Coefficient
MDS	Multi-Dimensional Scaling
NELL	Never-Ending Language Learning
NLP	Natural Language Processing
NP	Noun Phrase
OMCS	Open Mind Common Sense
OMR	Ontology for Media Resources
OWL	Ontology Web Language
PAM	Partitioning Around Medoids
PCA	Principal Component Analysis
POG	Prerequisite Outcome Goal
POS	Part of Speech
PP	Prepositional Phrase
PROM	Patient-Reported Outcome Measure
RDBMS	Relational Database Management Systems
RDF	Resource Description Framework
RDFS	Resource Description Framework Schema
RP	Random Projection
RPG	Role Play Game
RNN	Recursive Neural Network
RNTN	Recursive Neural Tensor Network
SKOS	Simple Knowledge Organization System
SQL	Structured Query Language
SRHT	Subsampled Randomized Hadamard Transform
SVD	Singular Value Decomposition
SVM	Support Vector Machine
TF-IDF	Term Frequency-Inverse Document Frequency
TMS	Truth Maintenance System
TSVD	Truncated Singular Value Decomposition
UI	User Interface
UGC	User-Generated Content
UML	Unified Modeling Language
XML	Extensible Markup Language
W3C	World Wide Web Consortium
WNA	WordNet-Affect

Chapter 1
Introduction

Everything we hear is an opinion, not a fact.
Everything we see is a perspective, not the truth.

Marcus Aurelius

Abstract This introductory chapter offers an updated literature review of sentiment analysis research and explains the importance of common-sense knowledge as a means to better understand natural language. In particular, the chapter proposes insights on the evolution of opinion mining research from heuristics to discourse structure, from coarse- to fine-grained analysis, and from keyword to concept-level polarity detection. Subsequently, a comprehensive literature review on common-sense knowledge representation is proposed, together with a discussion on why common-sense is important for sentiment analysis and natural language understanding. The chapter ends with an introduction of sentic computing as a new common-sense based framework for concept-level sentiment analysis and the explanation of its three key shifts, which highly differentiate it from standard approaches to opinion mining and social media analysis.

Keywords Opinion mining • Sentiment analysis • Sentic computing • Natural language processing • Common-sense knowledge

Between the year of birth of the Internet and 2003, the year of birth of social networks such as MySpace, Delicious, LinkedIn, and Facebook, there were just a few dozen exabytes of information on the Web. Today, that same amount of information is created weekly. The advent of the Social Web has provided people with new content-sharing services that allow them to create and share their own contents, ideas, and opinions, in a time- and cost-efficient way, with virtually millions of other people connected to the World Wide Web.

This huge amount of information, however, is mainly unstructured (because it is specifically produced for human consumption) and, hence, not directly machine-processable. The automatic analysis of text involves a deep understanding of natural language by machines, a reality from which we are still very far off. Hitherto, online information retrieval, aggregation, and processing have mainly been based

© Springer International Publishing Switzerland 2015

E. Cambria, A. Hussain, *Sentic Computing*, Socio-Affective Computing 1,
DOI 10.1007/978-3-319-23654-4_1

on algorithms relying on the textual representation of webpages. Such algorithms are very good at retrieving texts, splitting them into parts, checking spelling and counting the number of words. When it comes to interpreting sentences and extracting meaningful information, however, their capabilities are known to be very limited. Natural language processing (NLP), in fact, requires high-level symbolic capabilities [112], including:

- creation and propagation of dynamic bindings;
- manipulation of recursive, constituent structures;
- acquisition and access of lexical, semantic, and episodic memories;
- control of multiple learning/processing modules and routing of information among such modules;
- grounding of basic-level language constructs (e.g., objects and actions) in perceptual/motor experiences;
- representation of abstract concepts.

All such capabilities are required to shift from mere NLP to what is usually referred to as natural language understanding [11]. Today, most of the existing approaches are still based on the syntactic representation of text, a method that relies mainly on word co-occurrence frequencies. Such algorithms are limited by the fact that they can only process information they can 'see'. As human text processors, we do not have such limitations as every word we see activates a cascade of semantically related concepts, relevant episodes, and sensory experiences, all of which enable the completion of complex NLP tasks – such as word-sense disambiguation, textual entailment, and semantic role labeling – in a quick and effortless way.

Computational models attempt to bridge such a cognitive gap by emulating the way the human brain processes natural language, e.g., by leveraging on semantic features that are not explicitly expressed in text. Computational models are useful both for scientific purposes (such as exploring the nature of linguistic communication), as well as for practical purposes (such as enabling effective human-machine communication). Traditional research disciplines do not have the tools to completely address the complex intertwined problems of how language comprehension and production work. Even if you combine all the approaches, a comprehensive theory would be too complex to be studied using traditional methods. However, we may be able to realize such complex theories as computer programs and then test them by observing how well they perform. By seeing where they fail, we can incrementally improve them. Computational models may provide very specific predictions about human behaviors that can then be explored by the psycholinguist. By continuing this process, we may eventually acquire a deeper understanding of how human language processing occurs. To realize such a dream will take the combined efforts of forward-thinking multi-disciplinary teams of psycholinguists, neuroscientists, anthropologists, philosophers, and computer scientists.

This volume presents sentic computing as a new computational model at the crossroads between affective computing, information extraction, and common-sense reasoning, which exploits both computer and human sciences to better interpret and process social information on the Web. The structure of the volume is as follows: this chapter presents the state of the art of sentiment analysis research and common-sense computing, and introduces the three key shifts of sentic computing; Chap. 2 describes how SenticNet is built; Chap. 3 illustrates how SenticNet is used, in concomitance with linguistic patterns and machine learning, for polarity detection; Chap. 4 reports some recent literature on sentic computing and lists some applications of it; finally, Chap. 5 proposes concluding remarks and future work.

1.1 Opinion Mining and Sentiment Analysis

Sentiment-analysis systems can be broadly categorized into knowledge-based [63] or statistics-based systems [64]. While, initially, the use of knowledge bases was more popular for the identification of emotions and polarity in text, recently sentiment analysis researchers have been increasingly using statistics-based approaches, with a special focus on supervised statistical methods. For example, Pang et al. [238] compared the performance of different machine learning algorithms on a movie review dataset: using a large number of textual features they obtained 82.90 % of accuracy. A recent approach by Socher et al. [293] obtained even better accuracy (85 %) on the same dataset using a recursive neural tensor network (RNTN). Yu and Hatzivassiloglou [338] used semantic orientation of words to identify polarity at sentence level. Melville et al. [209] developed a framework that exploits word-class association information for domain-dependent sentiment analysis.

More recent studies such as [102, 171, 216], and [79], exploit microblogging text or Twitter-specific features such as emoticons, hashtags, URLs, @symbols, capitalizations, and elongations to enhance sentiment analysis of tweets. Tang et al. [304] developed a convolutional neural network based approach to obtain word embeddings for the words mostly used in tweets. These word vectors were then fed to a convolutional neural network for sentiment analysis. Santos et al. [277] also focused on deep convolutional neural network for sentiment detection in short text. Recent approaches also focus on developing word embeddings based on a sentiment corpora. Such word vectors called Sentiment Specific Word Embeddings [305] include more affective clues than regular word vectors and producing better result.

In alternative approaches, it is well known that many short n-grams are neutral while longer phrases are well distributed among positive and negative subjective sentence classes. Thus, matrix representations for long phrases and matrix multiplication to model composition are also being used to evaluate sentiment. In such models, sentence composition is modeled using deep neural networks such as recursive auto-associated memories [133, 162, 248]. Recursive neural networks (RNN) predict the sentiment class at each node in the parse tree and try to capture the negation and its scope in the entire sentence.

In the standard RNN, each word is represented as a vector and it is first determined which parent's children have already been computed. Next, the parent is computed via a composition function over child nodes. In Matrix RNN the composition function for long phrases depends on the words being combined and, hence, is linguistically motivated. However, the number of possible composition functions is exponential, hence in [293], a RNTN was introduced, which uses a single tensor composition function to define multiple bilinear dependencies between words. Most of the literature on sentiment analysis has focused on text written in English and consequently most of the resources developed, such as lexicons with sentiment labels, are in English. Adapting such resources to other languages can be considered as a domain adaptation problem [52, 334]. This section discusses the evolution of different approaches and depths of analysis [65], i.e., from heuristics to discourse structure (Sect. 1.1.1), from coarse- to fine-grained analysis (Sect. 1.1.2), from keyword to concept level opinion mining (Sect. 1.1.3).

1.1.1 From Heuristics to Discourse Structure

Several unsupervised learning approaches rely on the creation of a sentiment lexicon in an unsupervised manner that is later used to determine the degree of positivity (or subjectivity) of a text unit. The crucial component is, therefore, the creation of the lexicon via the unsupervised labeling of words or phrases with their sentiment polarity or subjectivity [237]. This lexicon can be used to identify the *prior polarity* or the *prior subjectivity* of terms or phrases, to use towards further identifying contextual polarity or subjectivity. Early works were mainly based on linguistic heuristics. For example, Hatzivassiloglou and McKeown's technique [140] was built on the fact that, in the case of polarity classification, the two classes of interest represent opposites, and 'opposition constraints' can be used to help label decisions.

Other works propagated the valence of seed words, for which the polarity is known, to terms that co-occur with them in general text or in dictionary glosses, or, to synonyms and words that co-occur with them in other WordNet-defined relations. A collective labeling approach can also be applied to opinion mining product features. Popescu and Etzioni [247] proposed an iterative algorithm that, starting from a global word label computed over a large collection of generic topic text, gradually tried to re-define such a label, first to one that is specific to a review corpus, then to one that is specific to a given product feature, and finally to one that is specific to the particular context in which the word occurs.

Further, Snyder and Barzilay [291] exploited the idea of utilizing discourse information to aid the inference of relationships between product attributes. They designed a linear classifier for predicting whether all aspects of a product are given the same rating, and combined such prediction with that of individual-aspect classifiers, in order to minimize a certain loss function. Regression techniques are often employed for the prediction of the degree of positivity in opinionated documents such as product reviews.

Regression enables implicit modeling of similarity relationships between classes that correspond to points on a scale, such as the number of 'stars' given by a reviewer [237]. Modeling discourse structure, such as twists and turns in documents, contributes to a more effective overall sentiment labeling. Early works attempted to partially address this problem via incorporating location information in the feature set [235]. More recent studies have underlined this position as particularly relevant in the context of sentiment summarization. In particular, in contrast to topic-based text summarization, where the incipits of articles usually serve as a strong baseline, the last n sentences of a review have been shown to serve as a much better summary of the overall sentiment of the document, and to be almost as good as the n (automatically-computed) most subjective sentences [235]. Joshi and Rose [161], for example, explored how features based on syntactic dependency relations can be utilized to improve performance in opinion mining. Using a transformation of dependency relation triples, they convert them into 'composite back-off features' that generalize better than the regular lexicalized dependency relation features.

1.1.2 From Coarse- to Fine-Grained

The evolution of research works in the field of opinion mining and sentiment analysis can be seen not only in the use of increasingly sophisticated techniques, but also in the different depths of analysis adopted. Early works aimed to classify entire documents as containing overall positive or negative polarity [239] or rating scores (e.g., 1–5 stars) of reviews [236]. These were mainly supervised approaches relying on manually labeled samples, such as movie or product reviews where the opinionist's overall positive or negative attitude was explicitly indicated. However, opinions and sentiments do not occur only at document level, nor are they limited to a single valence or target. Contrary or complementary attitudes toward the same topic or multiple topics can be present across the span of a document. Later works adopted a segment level opinion analysis aiming to distinguish sentimental from non-sentimental sections, e.g., by using graph-based techniques for segmenting sections of a document on the basis of their subjectivity [235], or by performing a classification based on some fixed syntactic phrases that are likely to be used to express opinions [310], or by bootstrapping using a small set of seed opinion words and a knowledge base such as WordNet [163].

In recent works, text analysis granularity has been taken down to sentence level, e.g., by using presence of opinion-bearing lexical items (single words or n-grams) to detect subjective sentences [168, 272], or by using semantic frames defined in FrameNet [19] for identifying the topics (or targets) of sentiment [169], or by exploiting association rule mining [4] for a feature-based analysis of product reviews [148]. Commonly, a certain degree of continuity exists in subjectivity labels of adjacent sentences, as an author usually does not switch too frequently between being subjective and being objective.

Hence, some works also propose a collective classification of the document based on assigning preferences for pairs of nearby sentences [236, 342]. All such approaches, however, are still some way from being able to infer the cognitive and affective information associated with natural language as they mainly rely on semantic knowledge bases which are still too limited to efficiently process text at sentence level. Moreover, such a text analysis granularity level might still not be enough as a single sentence may express more than one opinion [330].

1.1.3 From Keywords to Concepts

Existing approaches can be grouped into three main categories, with few exceptions: keyword spotting, lexical affinity, and statistical methods. Keyword spotting is the most naïve approach and probably also the most popular because of its accessibility and economy. Text is classified into affect categories based on the presence of fairly unambiguous affect words like 'happy', 'sad', 'afraid', and 'bored'. Elliott's Affective Reasoner [117], for example, searches for 198 affect keywords, e.g., 'distressed' and 'enraged', in addition to affect intensity modifiers, e.g., 'extremely', 'somewhat', and 'mildly', plus a handful of cue phrases, e.g., 'did that' and 'wanted to'.

Other popular sources of affect words are Ortony's Affective Lexicon [230], which groups terms into affective categories, and Wiebe's linguistic annotation scheme [328]. The weaknesses of this approach lie in two areas: (1) poor recognition of affect when negation is involved and (2) reliance on surface features. Regarding its first weakness, while the approach can correctly classify the sentence "today was a happy day" as being happy, it is likely to fail on a sentence like "today wasn't a happy day at all". In relation to its second weakness, the approach relies on the presence of obvious affect words which are only surface features of the prose.

In practice, a lot of sentences convey affect through underlying meaning rather than affect adjectives. For example, the text "My husband just filed for divorce and he wants to take custody of my children away from me" certainly evokes strong emotions, but uses no affect keywords, and therefore, cannot be classified using a keyword spotting approach. Lexical affinity is slightly more sophisticated than keyword spotting as, rather than simply detecting obvious affect words; it assigns arbitrary words a probabilistic 'affinity' for a particular emotion. For example, 'accident' might be assigned a 75 % probability of indicating a negative affect, as in 'car accident' or 'hurt by accident'. These probabilities are usually trained from linguistic corpora [264, 294, 300, 329].

Though often outperforming pure keyword spotting approaches, there are two main problems with the approach. First, lexical affinity, operating solely on the word level, can easily be tricked by sentences like "I avoided an accident" (negation) and "I met my girlfriend by accident" (other word senses). Second, lexical affinity probabilities are often biased towards text of a particular genre, dictated by the source of the linguistic corpora. This makes it difficult to develop a reusable, domain-independent model.

Statistical methods, such as latent semantic analysis (LSA) and support vector machine (SVM), have been popular for affect classification of texts and used by researchers on projects such as Goertzel's Webmind [134], Pang's movie review classifier [239], and many others [1, 107, 148, 227, 236, 311, 317]. By feeding a machine learning algorithm a large training corpus of affectively annotated texts, it is possible for the systems to not only learn the affective valence of affect keywords as in the keyword spotting approach, but such a system can also take into account the valence of other arbitrary keywords (like lexical affinity), punctuation, and word co-occurrence frequencies. However, statistical methods are generally considered to be semantically weak, that is, with the exception of obvious affect keywords, other lexical or co-occurrence elements in a statistical model have little predictive value individually. As a result, statistical text classifiers only work with acceptable accuracy when given a sufficiently large text input. So, while these methods may be able to affectively classify user's text at the page or paragraph level, they do not work well on smaller text units such as sentences.

1.2 Towards Machines with Common Sense

Communication is one of the most important aspects of human life. It is always associated with a cost in terms of energy and time, since information needs to be encoded, transmitted, and decoded, and, on occasions, such factors can even make the difference between life and death. This is why people, when communicating with each other, provide just the useful information and take the rest for granted. This 'taken for granted' information is what is termed 'common-sense' – obvious things people normally know and usually leave unstated. Common-sense is not the kind of knowledge we can find in Wikipedia,[1] rather it implicitly exists in all the basic relationships among words, concepts, phrases, and thoughts that allow people to communicate with each other and face everyday life problems. It is a kind of knowledge that sounds obvious and natural to us, but is actually daedal and multi-faceted.

The illusion of simplicity comes from the fact that, as each new group of skills matures, we build more layers on top of them and tend to forget about the previous layers. Common-sense, in fact, is not a simple thing. Instead, it is an immense society of hard-earned practical ideas, of multitudes of life-learned rules and exceptions, dispositions and tendencies, balances and checks [212]. This section discusses the importance of common-sense for the development of intelligent systems (Sect. 1.2.1) and illustrates different knowledge representation strategies (Sect. 1.2.2). The section also refers to a recently proposed survey on common-sense computing [54] to present the evolution of related research fields, from logic-based approaches to more recent methods based on natural language techniques (Sect. 1.2.3).

[1] http://wikipedia.org

1.2.1 The Importance of Common Sense

Concepts are the glue that holds our mental world together [221]. Without concepts, there would be no mental world in the first place [31]. Doubtless to say, the ability to organize knowledge into concepts is one of the defining characteristics of the human mind. Of the different sorts of semantic knowledge that is researched, arguably the most general and widely applicable kind is knowledge about the everyday world possessed by all people, what we refer to as common-sense knowledge. While to the average person the term common-sense is regarded as synonymous with good judgement, to the AI community it is used in a technical sense to refer to the millions of basic facts and understandings possessed by most people, e.g., "a lemon is sour", "to open a door, you must usually first turn the doorknob", "if you forget someone's birthday, they may be unhappy with you".

Common-sense knowledge, thus defined, spans a huge portion of human experience, encompassing knowledge about the spatial, physical, social, temporal, and psychological aspects of typical everyday life. Because it is assumed that every person possesses common-sense, such knowledge is typically omitted from social communications, such as text. A full understanding of any text then, requires a surprising amount of common-sense, which currently only people possess. Common-sense knowledge is what we learn and what we are taught about the world we live in during our formative years, in order to better understand and interact with the people and the things around us. Common-sense is not universal, rather cultural and context dependent. The importance of common-sense can be particularly appreciated when traveling to far away places, where sometimes it is necessary to almost entirely reset one's common-sense knowledge in order to more effectively integrate socially and intellectually.

Despite the language barrier, however, moving to a new place involves facing habits and situations that might go against what we consider basic rules of social interaction or things we were taught by our parents, such as eating with hands, eating from someone else's plate, slurping on noodle-like food or while drinking tea, eating on the street, crossing the road despite the heavy traffic, squatting when tired, removing shoes at home, growing long nails on your last fingers, or bargaining on anything you need to buy. This can happen also the other way round, that is, when you do something perfectly in line with your common-sense that violates the local norms, e.g., cheek kissing as a form of greeting.

Common-sense is the holistic knowledge (usually acquired in early stages of our lives) concerning all the social, political, economic, and environmental aspects of the society we live in. Machines, which have never had the chance to live a 'human-like' life, have no common-sense at all and, hence, know nothing about us. To help us work, computers must get to know what our jobs are. To entertain us, they need to know what we like and dislike. To take care of us, they have to know how we feel. To understand us, they must think as we think. Today, in fact, computers do only what they are programmed to do. They only have one way to deal with a problem and, if something goes wrong, usually get stuck. Nowadays we have computer programs that exceed the capabilities of world experts in certain problem-solving tasks, yet, as convincingly demonstrated by McClelland [207], are still not able to do what a 3 years old child can, at a range of simple cognitive tasks, such as object recognition, language comprehension, and planning and acting in contextually appropriate ways. It is because machines have no cognitive goals, no hopes, no fears; they do not know the meaning of life.

Computers can only do logical things, but meaning is an intuitive process – it cannot be simply reduced to zeros and ones. We will need to transmit to computers our common-sense knowledge of the world as there may actually not be enough capable human workers left to perform the necessary tasks for our rapidly ageing population. To deal with this emerging AI emergency,[2] we will be required to endow computers and machines with physical knowledge of how objects behave, social knowledge of how people interact, sensory knowledge of how things look and taste, psychological knowledge about the way people think, and so on. But having a simple database of millions of common-sense facts will not be enough: we will also have to teach computers how to handle and make sense of this knowledge, retrieve it when necessary, and contextually learn from experience – in a word, we will have to give them the capacity for common-sense reasoning.

1.2.2 Knowledge Representation

Since the very beginning, AI has rested on a foundation of formal representation of knowledge. Knowledge representation (KR) is a research area that directly addresses languages for representation and the inferences that go along with them. A central question in KR research relates to the form knowledge is best expressed in. One of the most popular representation strategies is first-order logic (FOL), a deductive system that consists of axioms and rules of inferences and can be used to formalize relationally rich predicates and quantification [24].

FOL supports syntax, semantics and, to a certain degree, pragmatics expressions. Syntax specifies the way groups of symbols are to be arranged, so that they can be considered properly formed. Semantics specify what well-formed expressions are supposed to mean. Pragmatics specifies how contextual information can be leveraged to provide better correlation between different semantics, for tasks such as word sense disambiguation. Logic, however, is known to have the problem of monotonicity. The set of entailed sentences can only increase as information is added to the knowledge base. This violates a common property of human reasoning, i.e., changing one's mind. Solutions such as default and linear logic serve to address parts of these issues. Default logic is proposed by Raymond Reiter to formalize default assumptions, e.g., "all birds fly" [268]. However, issues arise when default

[2]http://mitworld.mit.edu/video/484

logic formalize facts that are true in the majority of cases, but not always, e.g., "penguins do not fly".

Linear logic, or constructive logic, was developed by Arend Heyting [145]. It is a symbolic logical system that preserves justification, rather than truth, and supports rejecting the weakening and contraction rules. It excels in careful deductive reasoning and is suitable in situations that can be posed precisely. As long as a scenario is static and can be described in detail, situation-specific rules can perfectly model it but, when it comes to capture a dynamic and uncertain real-world environment, logical representation usually fails for lack of generalization capabilities. Accordingly, it is not natural for a human to encode knowledge in logical formalization. Another standard KR strategy, based on FOL, is the use of relational databases. The idea is to describe a database as a collection of predicates over a finite set of variables and describing constraints on the possible values. Structured query language (SQL) [100] is the database language designed for the retrieval and management of data in relational database management systems (RDBMS) [87]. Commercial (e.g., Oracle,[3] Sybase,[4] Microsoft SQL Server[5]) and open-source (e.g., mySQL[6]) implementations of RDBMS are available and commonly used in the IT industry.

Relational database design requires a strict process called normalization to ensure that the relational database is suitable for general purpose querying and the relational database is free of database operational anomalies. A minimal practical requirement is third normal form (3NF) [88], which is stricter than first and second normal forms and less strict as compared to Boyce-Codd normal form (BCNF) [89], fourth, and fifth normal forms. Stricter normal forms means that the database design is more structured and, hence, requires more database tables. The advantage is that the overall design looks more organized. The disadvantage is the performance trade-off when joint table SQL queries are invoked. Relational database design, moreover, does not directly address representation of parent-child relationships in the object-oriented paradigm, subjective degrees of confidence, and temporal dependent knowledge.

A popular KR strategy, especially among Semantic Web researchers, is production rule [82]. A production rule system keeps a working memory of on-going memory assertions. This working memory is volatile and maintains a set of production rules. A production rule comprises an antecedent set of conditions and a consequent set of actions (i.e., IF <conditions> THEN <actions>). The basic operation for a production rule system involves a cycle of three steps ('recognize', 'resolve conflict', and 'act') that repeats until no more rules are applicable to working memory. The step 'recognize' identifies the rules whose antecedent conditions are satisfied by the current working memory. The set of rules

[3]http://oracle.com

[4]http://sybase.com

[5]http://microsoft.com/sqlserver

[6]http://mysql.com

identified is also called the conflict set. The step 'resolve conflict' looks into the conflict set and selects a set of suitable rules to execute. The step 'act' simply executes the actions and updates the working memory. Production rules are modular. Each rule is independent from others, allowing rules to be added and deleted easily.

Production rule systems have a simple control structure and the rules are relatively easy for humans to understand. This is because rules are usually derived from observations of expert behavior or expert knowledge, thus the terminology used in encoding the rules tends to resonate with human understanding. However, there are issues with scalability when production rule systems grow larger. Significant maintenance overhead is required to maintain systems with thousands of rules.

Another prominent KR strategy among Semantic Web researchers is ontology web language (OWL),[7] an XML-based vocabulary that extends the resource description framework (RDF)[8] and resource description framework schema (RDFS)[9] to provide a more comprehensive ontology representation, such as the definition of classes, relationships between classes, properties of classes, and constraints on relationships between classes and properties of classes. RDF supports the subject-predicate-object model that makes assertion about a resource. Reasoning engines have been developed to check for semantic consistency and help improve ontology classification. OWL is a W3C recommended specification and comprises three dialects: OWL-Lite, OWL-DL, and OWL-Full. Each dialect comprises a different level of expressiveness and reasoning capabilities. OWL-Lite is the least expressive compared to OWL-Full and OWL-DL. It is suitable for building ontologies that only require classification hierarchy and simple constraints and, for this reason, provides the most computationally efficient reasoning capability. OWL-DL is more expressive than OWL-Full, but more expressive than OWL-Lite. It has restrictions on the use of some of the description tags, hence, computation performed by a reasoning engine on OWL-DL ontologies can be completed in a finite amount of time [174]. OWL-DL is so named due to its correspondence with description logic. It is also the most commonly used dialect for representing domain ontology for Semantic Web applications. OWL-Full is the complete language, and is useful for modeling a full representation of a domain. However, the trade-off for OWL-Full is the high complexity of the model that can result in sophisticated computations that may not complete in finite time. In general, OWL requires strict definition of static structures, hence, it is not suitable for representing knowledge that requires subjective degrees of confidence, but rather for representing declarative knowledge. OWL, moreover, does not allow easy representation of temporal dependent knowledge.

Another well-known way to represent knowledge is to use networks. Bayesian networks [244], for example, provide a means of expressing joint probability distributions over many interrelated hypotheses. Bayesian network is also called a belief network. All variables are represented using a directed acyclic graph

[7]http://w3.org/TR/owl-overview

[8]http://w3.org/TR/PR-rdf-syntax

[9]http://w3.org/2001/sw/wiki/RDFS

(DAG). The nodes of a DAG represent variables. Arcs are causal connections between two variables where the truth of the former directly affects the truth of the latter. A Bayesian network is able to represent subjective degrees of confidence. The representation explicitly explores the role of prior knowledge and combines evidence of the likelihood of events. In order to compute the joint distribution of the belief network, there is a need to know $\Pr(P|\text{parents}(P))$ for each variable P. It is difficult to determine the probability of each variable P in the belief network. Hence, it is also difficult to scale and maintain the statistical table for large scale information processing problems. Bayesian networks also have limited expressiveness, which is only equivalent to the expressiveness of proposition logic. For this reason, semantic networks are more often used for KR.

A semantic network [295] is a graphical notation for representing knowledge in patterns of interconnected nodes and arcs. There are six types of networks, namely definitional networks, assertional networks, implicational networks, executable networks, learning networks, and hybrid networks. A definitional network focuses on *IsA* relationships between a concept and a newly defined sub-type. The resulting network is called a generalization, which supports the rule of inheritance for copying properties defined for a super-type to all of its sub-types. Definitions are true by definition and, hence, the information in definitional networks is often assumed to be true. Assertional networks are meant to assert propositions and the information is assumed to be contingently true. Contingent truth means that the proposition is true in some but not in all the worlds. The proposition also has sufficient reason in which the reason entails the proposition, e.g., "the stone is warm" with the sufficient reasons being "the sun is shining on the stone" and "whatever the sun shines on is warm". Contingent truth is not the same as the truth that is assumed in default logic, rather it is closer to the truth assumed in model logic.

Implicational networks use implication as the primary relation for connecting nodes. They are used to represent patterns of beliefs, causality, or inferences. Methods for realizing implicational networks include Bayesian networks and logic inferences used in a truth maintenance system (TMS). By combinations of forward and backward reasoning, a TMS propagates truth-values to nodes whose truth-value is unknown.

Executable networks contains mechanisms implemented in run-time environment such as message passing, attached procedure (e.g., data-flow graph), and graph transformation that can cause change to the network. Learning networks acquire knowledge from examples by adding and deleting nodes and links, or by modifying weights associated with the links. Learning networks can be modified in three ways: rote memory, changing weights, and restructuring. As for the rote memory, the idea is to add information without making changes to the current network. Exemplar methods can be found in relational databases. For example, Patrick Winston used a version of relational graphs to describe structures, such as arches and towers [331]. When his program was given positive and negative examples of each type of structure, it would generalize the graphs to derive a definitional network for classifying all types of structures that were considered. The idea of changing weights, in turn, is to modify the weights of links without changing the network structure for the nodes and links. Exemplar methods can be found in neural networks.

As for restructuring, finally, the idea is to create fundamental changes to the network structure for creative learning. Methods include case-based reasoning, where the learning system uses rote memory to store various cases and associated actions such as the course of action. When a new case is encountered, the system finds those cases that are most similar to the new one and retrieves the outcome. To organize the search and evaluate similarity, the learning system must use restructuring to find common patterns in the individual cases and use those patterns as keys for indexing the database. Hybrid networks combine two or more of the previous techniques. Hybrid networks can be a single network, yet also comprise separate but closely interacting networks.

Sowa used unified modeling language (UML) as an example to illustrate a hybrid semantic network. Semantic networks are very expressive. The representation is flexible and can be used to express different paradigms such as relational models and hierarchical relationships. The challenge is at the implementation level. For example, it is difficult to implement a hybrid semantic network, which requires an integration of different methods.

1.2.3 Common-Sense Reasoning

What magical trick makes us intelligent? – Marvin Minsky was wondering more than two decades ago – The trick is that there is no trick. The power of intelligence stems from our vast diversity, not from any single, perfect principle [212]. The human brain is a very complex system, maybe the most complex in nature. The functions it performs are the product of thousands and thousands of different subsystems working together at the same time. Common-sense computing involves trying to emulate such mechanisms and, in particular, at exploiting common-sense knowledge to improve computers' understanding of the world. Before Minsky, many AI researchers started to think about the implementation of a common-sense reasoning based machine.

The very first person who seriously started thinking about the creation of such a machine was perhaps Alan Turing when, in 1950, he first raised the question "can machines think?". Whilst he never managed to answer that question, he provided the pioneering method to gauge artificial intelligence, the so called Turing test. The notion of common-sense in AI is actually dated 1958, when John McCarthy, in his seminal paper 'Programs with Common-Sense' [206], proposed a program, termed the 'advice taker', for solving problems by manipulating sentences in formal language. The main aim of such a program was to try to automatically deduce for itself a sufficiently wide class of immediate consequences of anything it was told and what it already knew. In his paper, McCarthy stressed the importance of finding a proper method of representing expressions in the computer since, according to him, in order for a program to be capable of learning something, it must first be capable of being told. He also developed the idea of creating a property list for each object, in which the specific things people usually know about that object are listed. It was the first attempt to build a common-sense knowledge base but, more importantly, it inspired the epiphany of the need for common sense to move forward in the technological evolution.

In 1959, McCarthy went to MIT and started, together with Minsky, the MIT Artificial Intelligence Project. They both were aware of the need for AI based on a common-sense reasoning approach, but while McCarthy was more concerned with establishing logical and mathematical foundations for it, Minsky was more involved with theories of how we actually reason using pattern recognition and analogy. These theories were organized some years later with the publication of the Society of Mind [212], a masterpiece of AI literature, which reveals an illuminating vision into how the human brain might work.

Minsky sees the mind made up of many little parts, termed 'agents', each mindless by itself but able to lead to true intelligence when working together. These groups of agents, called 'agencies', are responsible for performing some type of cognitive function, such as remembering, comparing, generalizing, exemplifying, analogizing, simplifying, predicting, and so on. The most common agents are the so called 'K-lines', whose task is simply to activate other agents: this is deemed to be a very important issue since agents are all highly interconnected and activating a K-line can cause a significant cascade of effects. To Minsky, mental activity is ultimately comprised of turning individual agents on and off: at any time only some agents are active and their combined activity constitutes the 'total state' of the mind. K-lines are a very simple but powerful mechanism since they allow entering a particular configuration of agents that formed a useful society in a past situation. This is how we build and retrieve cognitive problem solving strategies in our mind; and could also be how we ought to develop such problem solving strategies in our programs.

In 1990, McCarthy put together 17 papers to try to define common-sense knowledge by using mathematical logic in such a way that common-sense problems could be solved by logical reasoning. Deductive reasoning in mathematical logic has the so-called monotonicity property: if we add new assumptions to the set of initial assumptions, there may be some new conclusions, but every sentence that was a deductive consequence of the original hypotheses is still a consequence of the enlarged set.

Much of human reasoning is monotonic as well, but some important human common-sense reasoning is not. For example, if someone is asked to build a birdcage, the person may conclude that it is appropriate to put a top on it, but if one learns that the bird is in fact a penguin, such a conclusion may no longer be drawn. McCarthy formally described this assumption that things are as expected unless otherwise specified, with the 'circumscription method' of non-monotonic reasoning: a type of minimization similar to the closed world assumption that what is not known to be true is false. Around the same time, a similar attempt aimed at giving a shape to common-sense knowledge was reported by Ernest Davis [120]. He tried to develop an ad hoc language for expressing common-sense knowledge and inference techniques for carrying out common-sense reasoning in specific domains such as space, time, quantities, qualities, flows, goals, plans, needs, beliefs, intentions, actions, and interpersonal relations. Thanks to his and McCarthy's knowledge formalizations, the first steps were laid towards the expression of common-sense facts in a way that would have been suitable for inclusion in a general purpose database and, hence, towards the development of programs with common-sense.

Minsky's theory of human cognition, in particular, was welcomed with great enthusiasm by the AI community and gave birth to many attempts to build common-sense knowledge bases and develop systems capable of common-sense reasoning. The most representative projects are Cyc [186], Doug Lenat's logic-based repository of common-sense knowledge, WordNet [125], Christiane Fellbaum's universal database of word senses, and ThoughtTreasure [219], Erik Mueller's story understanding system. Cyc is one of the first attempts to assemble a massive knowledge base spanning human common-sense knowledge.

Initially started by Doug Lenat in 1984, this project utilizes knowledge engineers who hand-craft assertions and place them into a logical framework using CycL, Cyc's proprietary language. Cyc's knowledge is represented redundantly at two levels: a frame language distinction (epistemological level), adopted for its efficiency, and a predicate calculus representation (heuristic level), needed for its expressive power to represent constraints. While the first level keeps a copy of the facts in the uniform user language, the second level maintains its own copy in different languages and data structures suitable for manipulation by specialized inference engines. Knowledge in Cyc is also organized into 'microtheories', resembling Minsky's agencies, each with its own knowledge representation scheme and sets of assumptions. These microtheories are linked via 'lifting rules' that allow translation and communication of expressions between them.

Launched in 1985 at Princeton University, WordNet is a database of words (primarily nouns, verbs, and adjectives). It has been one of the most widely used resources in computational linguistics and text analysis primarily owing to the ease of interfacing it with any kind of application and system. The smallest unit in WordNet is the word/sense pair, identified by a 'sense key'. Word/sense pairs are linked by a small set of semantic relations such as synonyms, antonyms, *IsA* superclasses, and words connected by other relations such as *PartOf*. Each synonym set, in particular, is termed a 'synset': it comprises the representation of a concept, often explained through a brief gloss, and represents the basic building block for hierarchies and other conceptual structures in WordNet. Erik Mueller's ThoughtTreasure is a story understanding system with a great variety of common-sense knowledge on how to read and understand children's stories. It was inspired by Cyc and is similar to Cyc in that it has both natural language and common-sense components. But whereas Cyc mostly uses logic, ThoughtTreasure uses multiple representations schemes: grids for stereotypical settings, finite automata for rules of device behavior and mental processes, logical assertions for encyclopaedic facts and linguistic knowledge. ThoughtTreasure's lexicon is similar to WordNet but, while world knowledge is explicitly excluded from WordNet, ThoughtTreasure contains also concepts that are not lexicalized in English like 'going to the pub' or 'eating at the restaurant', which are very important for common-sense reasoning.

Using logic-based reasoning, in fact, can solve some problems in computer programming, but most real-world problems need methods better at matching patterns and constructing analogies, or making decisions based on previous experience with examples, or by generalizing from types of explanations that have worked well on similar problems in the past [213]. In building intelligent systems we have to try to reproduce our way of thinking: we turn ideas around in our mind to examine them from different perspectives until we find one that works for us. From this arises the need of using several representations, each integrated with its set of related pieces of knowledge, to be able to switch from one to another when one of them fails. The key, in fact, is using different representations to describe the same situation. Minsky blames our standard approach to writing a program for common-sense computing failures.

Since computers appeared, our approach to solve a problem has always consisted in first looking for the best way to represent the problem, and then looking for the best way to represent the knowledge needed to solve it and finally looking for the best procedure for solving it. This problem-solving approach is good when we have to deal with a specific problem, but there is something basically wrong with it: it leads us to write only specialized programs that cope with solving only that kind of problem. This is why, today, we have millions of expert programs but not even one that can be actually defined intelligent.

From here comes the idea of finding heterogeneous ways to represent common-sense knowledge and to link each unit of knowledge to the uses, goals, or functions that each knowledge-unit can serve. This non-monotonic approach reasserted by Minsky was adopted soon after by Push Singh within the Open Mind Common-Sense (OMCS) project [287]. Initially born from an idea of David Stork [301], the project differs from previous attempts to build a common-sense database for the innovative way to collect knowledge and represent it. OMCS is a second-generation common-sense database. Knowledge is represented in natural language, rather than using a formal logical structure, and information is not hand-crafted by expert engineers but spontaneously inserted by online volunteers. The reason why Lenat decided to develop an ad hoc language for Cyc is that vagueness and ambiguity pervade English and computer reasoning systems generally requiring knowledge to be expressed accurately and precisely. However, as expressed in the Society of Mind, ambiguity is unavoidable when trying to represent the common-sense world.

No single argument, in fact, is always completely reliable but, if we combine multiple types of arguments, we can improve the robustness of reasoning as well as improving the table stability by providing it with many small legs in place of just one very big leg. This way information is not only more reliable, but also stronger. If a piece of information goes lost, we can still access the whole meaning, exactly as the table keeps on standing up if we cut out one of the small legs. Diversity is, in fact, the key of OMCS' success: the problem is not choosing a representation in spite of another, but it is finding a way for them to work together in one system. The main difference between acquiring knowledge from the general public and acquiring it from expert engineers is that the general public is likely to leave as soon as they encounter something boring or difficult. The key is letting people do what they prefer to do. Different people, in fact, like to do different things: some like to enter new items, some like to evaluate items, others like to refine items. For this reason, OMCS is based on a distributed workflow model where the different stages of knowledge acquisition could be performed separately by different participants.

The system, in fact, was designed to allow users to insert new knowledge via both template-based input and free-form input, tag concepts, clarify properties, and validate assertions. But, since giving so much control to users can be dangerous, a fixed set of pre-validated sentences were meant to be presented to them from time to time, in order to assess their honesty, and the system was designed in a way that allowed users to reciprocally control each other by judging samples of each other's knowledge.

OMCS exploits a method termed cumulative analogy [81], a class of analogy-based reasoning algorithms that leverage existing knowledge to pose knowledge acquisition questions to the volunteer contributors. When acquiring knowledge online, the stickiness of the website is of primary importance. The best way to involve users in this case is by making them feel that they are contributing to the construction of a thinking machine and not just a static database. To do this, OMCS first determines what other topics are similar to the topic the user is currently inserting knowledge for, and then uses cumulative analogy to generate and present new specific questions about this topic.

1.3 Sentic Computing

With the dawn of the Internet Age, civilization has undergone profound, rapid-fire changes that we are experiencing more than ever today. Even technologies that are adapting, growing, and innovating have the gnawing sense that obsolescence is right around the corner. NLP research, in particular, has not evolved at the same pace as other technologies in the past 15 years.

While NLP research has made great strides in producing artificially intelligent behaviors, e.g., Google, IBM Watson, and Apple Siri, none of such NLP frameworks actually understand what they are doing – making them no different from a parrot that learns to repeat words without any clear understanding of what it is saying. Today, even the most popular NLP technologies view text analysis as a word- or pattern-matching task. Trying to ascertain the meaning of a piece of text by processing it at word level, however, is no different from attempting to understand a picture by analyzing it at pixel level.

In a Web, where 'Big Data' in the form of user-generated content (UGC) is drowning in its own output, NLP researchers are faced with the same challenge: the need to jump the curve [156] to make significant, discontinuous leaps in their thinking, whether it is about information retrieval, aggregation, or processing. Relying on arbitrary keywords, punctuation, and word co-occurrence frequencies

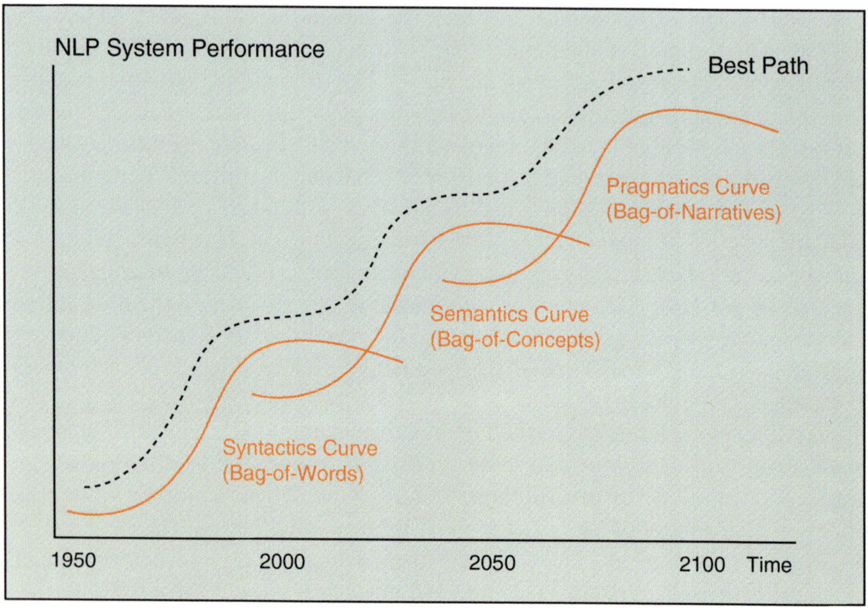

Fig. 1.1 Envisioned evolution of NLP research through three different eras or curves (Source: [67])

has worked fairly well so far, but the explosion of UGCs and the outbreak of deceptive phenomena such as web-trolling and opinion spam, are causing standard NLP algorithms to be increasing less efficient. In order to properly extract and manipulate text meanings, a NLP system must have access to a significant amount of knowledge about the world and the domain of discourse.

To this end, NLP systems will gradually stop relying too much on word-based techniques while starting to exploit semantics more consistently and, hence, make a leap from the Syntactics Curve to the Semantics Curve (Fig. 1.1). NLP research has been interspersed with word-level approaches because, at first glance, the most basic unit of linguistic structure appears to be the word. Single-word expressions, however, are just a subset of concepts, multi-word expressions that carry specific *semantics and sentics* [50], that is, the denotative and connotative information commonly associated with real-world objects, actions, events, and people.

Sentics, in particular, specifies the affective information associated with such real-world entities, which is key for common-sense reasoning and decision-making. Semantics and sentics include common-sense knowledge (which humans normally acquire during the formative years of their lives) and common knowledge (which people continue to accrue in their everyday life) in a re-usable knowledge base for machines. Common knowledge includes general knowledge about the world, e.g., *a chair is a type of furniture*, while common-sense knowledge comprises obvious or widely accepted things that people normally know about the world but which are

Fig. 1.2 A 'pipe' is not a pipe, unless we know how to use it (Source: [67])

usually left unstated in discourse, e.g., that *things fall downwards (and not upwards)* and *people smile when they are happy*.

The difference between common and common-sense knowledge can be expressed as the difference between knowing the name of an object and understanding the same object's purpose. For example, you can know the name of all the different kinds or brands of 'pipe', but not its purpose nor the method of usage. In other words, a 'pipe' is not a pipe unless it can be used [201] (Fig. 1.2).

It is through the combined use of common and common-sense knowledge that we can have a grip on both high- and low-level concepts as well as nuances in natural language understanding and therefore effectively communicate with other people without having to continuously ask for definitions and explanations.

Common-sense, in particular, is key in properly deconstructing natural language text into sentiments according to different contexts – for example, in appraising the concept `small_room` as negative for a hotel review and `small_queue` as positive for a post office, or the concept `go_read_the_book` as positive for a book review but negative for a movie review. Semantics, however, is just one layer up in the scale that separates NLP from natural language understanding. In order to achieve the ability to accurately and sensibly process information, computational models will also need to be able to project semantics and sentics in time, compare them in a parallel and dynamic way, according to different contexts and with respect to different actors and their intentions [147]. This will mean jumping from the Semantics Curve to the Pragmatics Curve, which will enable NLP to be more adaptive and, hence, open-domain, context-aware, and intent-driven. Intent, in particular, will be key for tasks such as sentiment analysis – a concept that generally has a negative connotation, e.g., small_seat, might turn out to be positive, e.g., if the intent is for an infant to be safely seated in it.

While the paradigm of the Syntactics Curve is the bag-of-words model [340] and the Semantics Curve is characterized by a bag-of-concepts model [50], the paradigm of the Pragmatics Curve will be the bag-of-narratives model. In this last model, each piece of text will be represented by mini-stories or interconnected episodes, leading to a more detailed level of text comprehension and sensible computation. While the bag-of-concepts model helps to overcome problems such as word-sense disambiguation and semantic role labeling, the bag-of-narratives model will enable tackling NLP issues such as co-reference resolution and textual entailment.

Sentic computing is a multi-disciplinary approach to natural language processing and understanding that represents a preliminary attempt to jump from the Semantics Curve to the Pragmatics Curve. By stepping away from the blind use of word co-occurrence frequencies and by working at concept level (Chap. 2), sentic computing already implemented the leap from the Syntactics Curve to the Semantics Curve. Through the introduction of linguistic patterns (Chap. 3), sentic computing is now gradually shifting to phrase structure understanding and narrative modeling.

In sentic computing, whose term derives from the Latin 'sentire' (root of words such as sentiment and sentience) and 'sensus' (as in common-sense), the analysis of natural language is based on common-sense reasoning tools, which enable the analysis of text not only at document, page or paragraph level, but also at sentence, clause, and concept level. Sentic computing is very different from common methods for polarity detection as it takes a multi-faceted approach to the problem of sentiment analysis. Some of the most popular techniques for opinion mining simply focus on word co-occurrence frequencies and statistical polarity associated with words. Such approaches can correctly infer the polarity of unambiguous text with simple phrase structure and in a specific domain (i.e., the one the statistical classifier has been trained with). One of the main characteristics of natural language, however, is ambiguity. A word like big does not really hold any polarity on its own as it can either be negative, e.g., in the case of big_problem, or positive, e.g., in big_meal, but most statistical methods assign a positive polarity to it, as this often appears in a positive context.

By working at concept level, sentic computing overcomes this and many other common problems of opinion-mining frameworks that heavily rely on statistical properties of words. In particular, sentic computing novelty gravitates around three key shifts:

1. the shift from a mere computer-science methodology to a multi-disciplinary approach to sentiment analysis (Sect. 1.3.1);
2. the shift from word-based text processing to the concept-level analysis of natural language sentences (Sect. 1.3.2);
3. the shift from the blind use of statistical properties to the ensemble application of common-sense knowledge and linguistic patterns (Sect. 1.3.3).

1.3.1 From Mono- to Multi-Disciplinarity

Sentic computing proposes the ensemble application of AI and Semantic Web techniques, for knowledge representation and inference; mathematics, for carrying out tasks such as graph mining and multi-dimensionality reduction; linguistics, for discourse analysis and pragmatics; psychology, for cognitive and affective modeling; sociology, for understanding social network dynamics and social influence; finally ethics, for understanding related issues about the nature of the mind and the creation of emotional machines.

In this volume, this shift will be illustrated through the integration of the Hourglass Model (Sect. 2.3.2), a biologically-inspired and psychologically-motivated emotion categorization model, with Sentic Neurons (Sect. 2.3.3), a novel classification framework based on artificial neural networks, to generate *sentics*, i.e., the connotative or affective information associated with natural language concepts.

1.3.2 From Syntax to Semantics

Sentic computing adopts the bag-of-concepts model in stead of simply counting word co-occurrence frequencies in text. Working at concept level entails preserving the meaning carried by multi-word expressions such as `cloud_computing`, which represent 'semantic atoms' that should never be broken down into single words. In the bag-of-words model, for example, the concept `cloud_computing` would be split into `computing` and `cloud`, which may wrongly activate concepts related to the weather and, hence, compromise categorization accuracy.

In this volume, this shift will be illustrated through sentic activation (Sect. 2.3.1), a novel framework for parallel analogy that applies an ensemble of spreading activation and multi-dimensional scaling to generate *semantics*, i.e., the denotative or conceptual information associated with natural language concepts.

1.3.3 From Statistics to Linguistics

Sentic computing allows sentiments to flow from concept to concept based on the dependency relation between clauses. The sentence "iPhone6 is expensive but nice", for example, is equal to "iPhone6 is nice but expensive" from a bag-of-words perspective. However, the two sentences bear opposite polarity: the former is positive as the user seems to be willing to make the effort to buy the product despite its high price, the latter is negative as the user complains about the price of iPhone6 although he/she likes it.

In this volume, this shift will be illustrated through the application of sentic patterns (Sect. 3), a powerful set of linguistic patterns to be used in concomitance with SenticNet (Sect. 2) for the sentiment analysis task of polarity detection.

Chapter 2
SenticNet

Where there is no love, there is no understanding.

Oscar Wilde

Abstract SenticNet is the knowledge base which the sentic computing framework leverages on for concept-level sentiment analysis. This chapter illustrates how such a resource is built. In particular, the chapter thoroughly explains the processes of knowledge acquisition, representation, and reasoning, which contribute to the generation of semantics and sentics that form SenticNet. The first part consists of a description of the knowledge sources used. The second part of the chapter illustrates how the collected knowledge is merged and represented redundantly at three levels: semantic network, matrix, and vector space. Finally, the third part presents the graph-mining and dimensionality-reduction techniques used to perform analogical reasoning, emotion recognition, and polarity detection.

Keywords Knowledge representation and reasoning • Semantic network • Vector space model • Spreading activation • Emotion categorization

SenticNet is a publicly available semantic resource for concept-level sentiment analysis that exploits an ensemble of graph mining and multi-dimensional scaling to bridge the conceptual and affective gap between word-level natural language data and the concept-level opinions and sentiments conveyed by them [61]. SenticNet is a knowledge base that can be employed for development of applications in diverse fields such as big social data analysis, human-computer interaction, and e-health. It is available either as a standalone XML repository[1] or as an API.[2]

SenticNet provides the semantics and sentics associated with 30,000 common-sense concepts, instantiated by either single words or multi-word expressions. A full list of such concepts is available at *http://sentic.net/api/en/concept*. Other API

[1] http://sentic.net/senticnet-3.0.zip

[2] http://sentic.net/api

© Springer International Publishing Switzerland 2015

E. Cambria, A. Hussain, *Sentic Computing*, Socio-Affective Computing 1,
DOI 10.1007/978-3-319-23654-4_2

```xml
<?xml version="1.0" encoding="UTF-8"?>
<rdf:RDF xmlns:rdf="http://w3.org/1999/02/22-rdf-syntax-ns#">
    <rdf:Description rdf:about="http://sentic.net/api/en/concept/celebrate_special_occasion">
        <rdf:type rdf:resource="http://sentic.net/api/concept"/>
        <text xmlns="http://sentic.net">celebrate special occasion</text>
        <semantics rdf:resource="http://sentic.net/api/en/concept/celebrate_holiday"/>
        <semantics rdf:resource="http://sentic.net/api/en/concept/celebrate_occasion"/>
        <semantics rdf:resource="http://sentic.net/api/en/concept/celebrate_birthday"/>
        <semantics rdf:resource="http://sentic.net/api/en/concept/celebrate_wedding"/>
        <semantics rdf:resource="http://sentic.net/api/en/concept/express_appreciation"/>
        <pleasantness rdf:datatype="http://w3.org/2001/XMLSchema#float">0.93</pleasantness>
        <attention rdf:datatype="http://w3.org/2001/XMLSchema#float">0.724</attention>
        <sensitivity rdf:datatype="http://w3.org/2001/XMLSchema#float">0</sensitivity>
        <aptitude rdf:datatype="http://w3.org/2001/XMLSchema#float">0</aptitude>
        <polarity rdf:datatype="http://w3.org/2001/XMLSchema#float">0.551</polarity>
    </rdf:Description>
</rdf:RDF>
```


Fig. 2.1 Sentic API concept call sample (Source: http://sentic.net/api)

methods include *http://sentic.net/api/en/concept/CONCEPT_NAME*, to retrieve all the available information associated with a specific concept, and more fine-grained methods to get semantics, sentics, and polarity, respectively:

1. *http://sentic.net/api/en/concept/CONCEPT_NAME/semantics*
2. *http://sentic.net/api/en/concept/CONCEPT_NAME/sentics*
3. *http://sentic.net/api/en/concept/CONCEPT_NAME/polarity*

In particular, the first command returns five SenticNet entries that are semantically related to the input concept, the second provides four affective values in terms of the dimensions of the Hourglass of Emotions (Sect. 2.3.2), and the third returns a float number between −1 and 1, which is calculated in terms of the sentics and specifies if (and to which extent) the input concept is positive or negative. For example, the full set of conceptual features associated with the multi-word expression `celebrate_special_occasion` can be retrieved with the following API call (Fig. 2.1): *http://sentic.net/api/en/concept/celebrate_special_occasion*

In case only the semantics associated with `celebrate_special_occasion` are needed, e.g., for gisting or auto-categorization tasks, they can be obtained by simply appending the command *semantics* to the above (Fig. 2.2). Similarly, the sentics associated with `celebrate_special_occasion`, useful for tasks such as affective HCI or theory of mind, can be retrieved by adding the command *sentics* (Fig. 2.3). Sentics can be converted to emotion labels, e.g., 'joy' and 'anticipation' in this case, by using the Hourglass model.

Finally, the polarity associated with `celebrate_special_occasion`, which can be exploited for more standard sentiment-analysis tasks, can be obtained through the command *polarity* (Fig. 2.4).

Unlike many other sentiment-analysis resources, SenticNet is not built by manually labeling pieces of knowledge coming from general NLP resources such as WordNet or DBPedia. Instead, it is automatically constructed by applying graph-mining and dimensionality-reduction techniques on the affective

```
<?xml version="1.0" encoding="UTF-8"?>
<rdf:RDF xmlns:rdf="http://w3.org/1999/02/22-rdf-syntax-ns#">
    <rdf:Description rdf:about="http://sentic.net/api/en/concept/celebrate_special_occasion/semantics">
        <rdf:type rdf:resource="http://sentic.net/api/concept/semantics"/>
        <semantics rdf:resource="http://sentic.net/api/en/concept/celebrate_holiday"/>
        <semantics rdf:resource="http://sentic.net/api/en/concept/celebrate_occasion"/>
        <semantics rdf:resource="http://sentic.net/api/en/concept/celebrate_birthday"/>
        <semantics rdf:resource="http://sentic.net/api/en/concept/celebrate_wedding"/>
        <semantics rdf:resource="http://sentic.net/api/en/concept/express_appreciation"/>
    </rdf:Description>
</rdf:RDF>
```


Fig. 2.2 Sentic API concept semantics call (Source: http://sentic.net/api)


```
<?xml version="1.0" encoding="UTF-8"?>
<rdf:RDF xmlns:rdf="http://w3.org/1999/02/22-rdf-syntax-ns#">
    <rdf:Description rdf:about="http://sentic.net/api/en/concept/celebrate_special_occasion/sentics">
        <rdf:type rdf:resource="http://sentic.net/api/concept/sentics"/>
        <pleasantness rdf:datatype="http://w3.org/2001/XMLSchema#float">0.93</pleasantness>
        <attention rdf:datatype="http://w3.org/2001/XMLSchema#float">0.724</attention>
        <sensitivity rdf:datatype="http://w3.org/2001/XMLSchema#float">0</sensitivity>
        <aptitude rdf:datatype="http://w3.org/2001/XMLSchema#float">0</aptitude>
    </rdf:Description>
</rdf:RDF>
```


Fig. 2.3 Sentic API concept sentics call (Source: http://sentic.net/api)


```
<?xml version="1.0" encoding="UTF-8"?>
<rdf:RDF xmlns:rdf="http://w3.org/1999/02/22-rdf-syntax-ns#">
    <rdf:Description rdf:about="http://sentic.net/api/en/concept/celebrate_special_occasion/polarity">
        <rdf:type rdf:resource="http://sentic.net/api/concept/polarity"/>
        <polarity rdf:datatype="http://w3.org/2001/XMLSchema#float">0.551</polarity>
    </rdf:Description>
</rdf:RDF>
```


Fig. 2.4 Sentic API concept polarity call (Source: http://sentic.net/api)

common-sense knowledge collected from three different sources (Sect. 2.1). This knowledge is represented redundantly at three levels: semantic network, matrix, and vector space (Sect. 2.2). Subsequently, semantics and sentics are calculated though the ensemble application of spreading activation, neural networks and an emotion categorization model (Sect. 2.3). The SenticNet construction framework (Fig. 2.5) merges all these techniques and models together in order to generate a knowledge base of 30,000 concepts and a set of semantics, sentics, and polarity for each.

2.1 Knowledge Acquisition

This section describes the knowledge bases and knowledge sources SenticNet is built upon. SenticNet mainly leverages on the general common-sense knowledge extracted from the Open Mind Common Sense initiative (Sect. 2.1.1), the affective knowledge coming from WordNet-Affect (Sect. 2.1.2) and the practical common-sense knowledge crowdsourced from GECKA (Sect. 2.1.3).

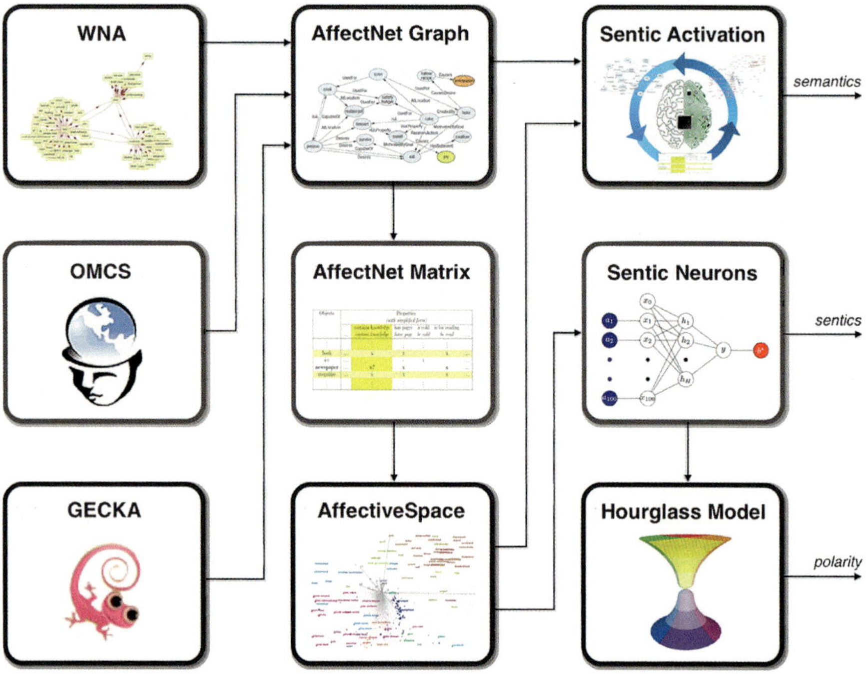

Fig. 2.5 SenticNet construction framework: by leveraging on an ensemble of graph mining and multi-dimensional scaling, this framework generates the semantics and sentics that form the SenticNet knowledge base (Source: The Authors)

2.1.1 Open Mind Common Sense

Open Mind Common Sense (OMCS) is an artificial intelligence project based at the MIT Media Lab whose goal is to build and utilize a large common-sense knowledge base from the contributions of many thousands of people across the Web.

Since its launch in 1999, it has accumulated more than a million English facts from over 15,000 contributors, in addition to leading to the development of knowledge bases in other languages. The project was the brainchild of Marvin Minsky, Push Singh, and Catherine Havasi. Development work began in September 1999, and the project was opened to the Internet a year later. Havasi described it in her dissertation as "an attempt to . . . harness some of the distributed human computing power of the Internet, an idea which was then only in its early stages" [141]. The original OMCS was influenced by the website Everything2, a collaborative Web-based community consisting of a database of interlinked user-submitted written material, and presented a minimalist interface that was inspired by Google.

There are many different types of knowledge in OMCS. Some statements convey relationships between objects or events, expressed as simple phrases of natural language: some examples include "A coat is used for keeping warm", "The sun is very hot", and "The last thing you do when you cook dinner is wash your dishes". The database also contains information on the emotional content of situations, in such statements as "Spending time with friends causes happiness" and "Getting into a car wreck makes one angry". OMCS contains information on people's desires and goals, both large and small, such as "People want to be respected" and "People want good coffee" [298]. Originally, these statements could be entered into the Web site as unconstrained sentences of text, which had to be parsed later. The current version of the Web site collects knowledge only using more structured fill-in-the-blank templates. OMCS also makes use of data collected by the Game With a Purpose "Verbosity" [8].

OMCS differs from Cyc because it has focused on representing the commonsense knowledge it collected as English sentences, rather than using a formal logical structure. Due to its emphasis on informal conceptual-connectedness over formal linguistic-rigor, OMCS knowledge is structured more like WordNet than Cyc. In its native form, the OMCS database is simply a collection of these short sentences that convey some common knowledge. In order to use this knowledge computationally, it has to be transformed into a more structured representation.

2.1.2 WordNet-Affect

WordNet-Affect (WNA) [302] is an extension of WordNet Domains, including a subset of synsets suitable to represent affective concepts correlated with affective words. Similar to the method used for domain labels, a number of WordNet synsets is assigned to one or more affective labels (a-labels). In particular, the affective concepts representing emotional state are individuated by synsets marked with the a-label emotion. There are also other a-labels for those concepts representing moods, situations eliciting emotions, or emotional responses. The resource was extended with a set of additional a-labels (called emotional categories), hierarchically organized, in order to specialize synsets with a-label emotion. The hierarchical structure of new a-labels was modeled on the WordNet hyperonym relation.

In the second stage, some modifications were introduced, in order to distinguish synsets according to emotional valence. Four additional a-labels were defined: positive, negative, ambiguous, and neutral (Table 2.1). The first one corresponds to positive emotions, defined as emotional states characterized by the presence of positive edonic signals (or pleasure). It includes synsets such as joy#1 or enthusiasm#1. Similarly the negative a-label identifies negative emotions characterized by negative edonic signals (or pain), for example anger#1 or sadness#1. Synsets representing affective states whose valence depends on semantic context (e.g., surprise#1) were marked with the tag ambiguous. Finally, synsets referring to mental states that are generally considered affective but are not characterized by valence, were marked with the tag neutral.

Another important property for affective lexicon concerning mainly adjectival interpretation is the stative/causative dimension. An emotional adjective is said

Table 2.1 A-Labels and corresponding example synsets (Source: http://wndomains.fbk.eu/wnaffect.html)

A-Labels	Examples
Emotion	Noun anger#1, verb fear#1
Mood	Noun animosisy#1, adjective amiable#1
Trait	Noun aggressiveness#1, adjective competitive#1
Cognitive state	Noun confusion#2, adjective dazed#2
Physical state	Noun illness#1, adjective all in#1
Hedonic signal	Noun hurt#3, noun suffering#4
Emotion-eliciting situation	Noun awkwardness#3, adjective out of danger#1
Emotional response	Noun cold sweat#1, verb tremble#2
Behavior	Noun offense#1, adjective inhibited#1
Attitude	Noun intolerance#1, noun defensive#1
Sensation	Noun coldness#1, verb feel#3

causative if it refers to some emotion that is caused by the entity represented by the modified noun (e.g., amusing movie). In a similar way, an emotional adjective is said stative if it refers to the emotion owned or felt by the subject denoted by the modified noun (e.g., cheerful/happy boy).

All words can potentially convey affective meaning. Each of them, even those more apparently neutral, can evoke pleasant or painful experiences. While some words have emotional meaning with respect to the individual story, for many others the affective power is part of the collective imagination (e.g., words such as mum, ghost, war etc.). Therefore, it is interesting to individuate a way to measure the affective meaning of a generic term. To this aim, the use of words in textual productions was studied, and in particular their co-occurrences with the words in which the affective meaning is explicit. We aim to distinguish between words directly referring to emotional states (e.g., fear, cheerful) and those having only an indirect reference that depends on the context (e.g., words that indicate possible emotional causes as monster or emotional responses as cry). The former are termed 'direct affective words' and the latter 'indirect affective words'.

Direct affective words were first integrated in WNA; then, a selection function (named Affective-Weight) based on a semantic similarity mechanism automatically acquired in an unsupervised way from a large corpus of texts (100 millions of words) was applied in order to individuate the indirect affective lexicon. Applied to a concept (e.g., a WordNet synset) and an emotional category, this function returns a value representing the semantic affinity with that emotion. In this way it is possible to assign a value to the concept with respect to each emotional category, and eventually select the emotion with the highest value. Applied to a set of concepts that are semantically similar, this function selects subsets characterized by some given affective constraints (e.g., referring to a particular emotional category or valence).

Authors were able to focus selectively on positive, negative, ambiguous or neutral types of emotions. For example, given difficulty as an input term, the system suggests as related emotions (a-labels): identification, negative-concern, ambiguous-expectation, and apathy. Moreover, given an input word (e.g., university) and the indication of an emotional valence (e.g., positive), the system suggests a set of related words through some positive emotional category (e.g., professor, scholarship, achievement) found through the emotions enthusiasm, sympathy, devotion, encouragement. These fine-grained kinds of affective lexicon selections can open up new possibilities in many applications that exploit verbal communication of emotions.

2.1.3 GECKA

Games with a purpose (GWAPs) are a simple yet powerful means to collect useful information from players in a way that is entertaining for them. Over the past few years, GWAPs have sought to exploit the brainpower made available by multitudes of casual gamers to perform tasks that, despite being relatively easy for humans to complete, are rather unfeasible for machines. The key idea is to integrate tasks such as image tagging, video annotation, and text classification into games, [5] producing win-win situations where people have fun while actually doing something useful. These games focus on exploiting player input to (syntax, not: both create) create both meaningful data and provide more enjoyable game experiences [306]. The problem with current GWAPs is that information gathered from them is often unrecyclable; acquired data is often applicable only to the specific stimuli encountered during gameplay. Moreover, such games often have a fairly low 'sticky factor', and are often unable to engage gamers for more than a few minutes.

The game engine for common-sense knowledge acquisition (GECKA) [62] implements a new GWAP concept that aims to overcome the main drawbacks of traditional data-collecting games by empowering users to create their own GWAPs and by mining knowledge that is highly reusable and multi-purpose. In particular, GECKA allows users to design compelling serious games for their peers to play while gathering common-sense knowledge useful for intelligent applications in any field requiring in-depth knowledge of the real world, including reasoning, perception and social systems simulation.

In addition to allowing for the acquisition of knowledge from game designers, GECKA enables players of the finished games to be educated in useful ways, all while being entertained. The knowledge gained from GECKA is later encoded in AffecNet in the form <concept-relationship-concept>. The use of this natural language based (rather than logic-based) framework allows GECKA players to conceptualize the world in their own terms, at a personalized level of semantic abstraction. Players can work with knowledge exactly as they envisage it, and researchers can access data on the same level as players' thoughts, significantly enhancing the usefulness of the captured data.

2.1.3.1 GWAP

GWAPs are an example of an emerging class of games that can be considered 'human algorithms', since humans act as processing nodes for problems that computers cannot yet solve. By providing an incentive for players, GWAPs gain a large quantity of computing power that can be harnessed for multiple applications, e.g., content tagging, ontology building, and knowledge acquisition by the general public.

GWAPs are possibly most well-known for image annotation. In the 'ESP' game [6], for example, players guess content objects or properties of random images by typing what they see when it appears on the screen. Other image annotation games include: Matchin [138], which focuses on perceived image quality by asking players to choose, in a pairwise manner, the picture they like better, and Phetch [7], a game that collects explanatory descriptions of images in order to improve Web accessibility for the visually impaired. Peekaboom [9] focuses on locating objects within images by letting a player select and reveal specific parts of an image and then challenging the other to guess the correct object name, while Squigl challenges players to spot objects in images previously annotated within the ESP Game. 'Picture This' requires players to choose from a set of images the one that best suits the given query. Image annotation games also include those intended to help streamline the robustness of CAPTCHAs, such as Magic Bullet [336], a team game in which players need to agree on the meaning of CAPTCHAs, and TagCaptcha [218], where players are asked to quickly describe CAPTCHA images with single words.

Besides images, GWAPs have been used for video annotation. For example, OntoTube [288], Yahoo's Videotaggame [343], and Waisd [3], are all games in which two players have to quickly agree on a set of tags for the same streaming YouTube video. GWAPs have also been exploited to automatically tag music tracks with semantic labels. HerdIt [23], for example, asks players to accomplish various tasks and answer quizzes related to the song they are listening to. In Tagatune [180], two players listen to an audio file and describe to the other what they are hearing. Players must then decide whether or not the game has played the same soundtrack to both participants. Sophisticated GWAPs have also attempted to perform complex tasks such as Web-page annotation and ontology building. Page Hunt [198], for example, is a GWAP that shows players Web pages and asks the user to guess what queries would generate those pages within the top 5 hits.

Results are used to improve the Microsoft Bing search engine. The game then shows players the top five page hits for the entered keywords and rewards are granted depending on how highly-ranked the assigned Web pages are within the result set. Another example, OntoPronto [288], is a quiz game for vocabulary building that attempts to build a large domain ontology from Wikipedia articles. Players receive random articles, which they map to the most specific and appropriate class of the Proton ontology (using the *subClassOf* relationship).

Another interesting game for generating domain ontologies from open data is Guess What?! [204]. Given a seed concept, a player has to find the matching URI in

DBpedia, Freebase and OpenCyc. The resulting labels/URIs are analyzed by simple computer-game-design tools in order to identify expressions that can be translated into logical operators, breaking down complex descriptions into small fragments. The game starts with the most general fragment and, at each round, a more specific fragment is connected to it through a logical operator, with players having to guess the concept described. Other GWAPs aim to align ontologies. Wordhunger, for example, is a Web-based application mapping WordNet synsets to Freebase. Each game round consists of a WordNet term and up to three suggested possible Freebase articles, among which players have to select the most fitting.

SpotTheLink is a two player game focusing on the alignment of random concepts from the DBpedia Ontology to the Proton upper ontology. Each player has to select Proton concepts that are either the same as, or, more specific than a randomly selected DBpedia concept. Data generated by SpotTheLink generates a SKOS mapping between the concepts of the two input ontologies. Finally, Wikiracing, Wiki Game, Wikispeedia and WikipediaMaze are games which aim to improve Wikipedia by engaging gamers in finding connections between articles by clicking links within article texts. WikipediaGame and Wikispedia focus on completing the race faster and with fewer clicks than other players. On the other hand, WikipediaMaze allows players to create races for each other and are incentivized to create and play races through the possibility of earning badges.

One of the most interesting tasks GWAPs can be used for is common-sense knowledge acquisition from members of the general public. One example, Verbosity [8], is a real time quiz game for collecting common-sense facts. In the game, two players take different roles at different times: one functions as a narrator, who has to describe a word using templates, while the other has to guess the word in the shortest time possible. FACTory Game [186] is a GWAP developed by Cycorp which randomly chooses facts from Cyc and presents them to players in order for them to guess whether a statement is true, false, or does not make sense. A variant of the FACTory game is the Concept Game on Facebook [144], which collects common-sense knowledge by proposing random assertions to users (along the lines of a slot machine) and gets them to decide whether the given assertion is meaningful or not. Virtual Pet [173] aims to construct a semantic network that encodes common-sense knowledge, and is built upon PPT, a popular Chinese bulletin board system accessible through a terminal interface. In this game each player owns a pet, which they take care of by asking and answering questions.

The pet acts as a stand-in for other players who then receive these questions and answers, and have to respond to, or validate them. Similar to Virtual Pet, the Rapport Game [173] draws on player efforts in constructing a semantic network that encodes common-sense knowledge. The Rapport Game, however, is built on top of Facebook and uses direct interaction between players. Finally, the Hourglass Game [68] is a timed game that associates natural language concepts with affective labels on a hourglass-shaped emotion categorization model. Players not only earn points in accordance with the accuracy of their associations, but also for their speed in creating affective matches. The game is able to collect new pieces of affective common-sense knowledge by randomly proposing multi-word expressions for which no affective information is known. The aggregation of this information generates a list of affective common-sense concepts, each weighted by a confidence score proportional to an inter-annotator agreement, which is therefore highly useful for opinion mining and sentiment analysis.

2.1.3.2 GECKA Key Functionalities

An important difference between traditional artificial intelligence (AI) systems and human intelligence is the human ability to harness common-sense knowledge gleaned from a lifetime of learning and experience to make informed decisions. This allows humans to adapt easily to novel situations where AI fails catastrophically due to a lack of situation-specific rules and generalization capabilities. Common-sense knowledge also provides background information enabling humans to successfully operate in social situations where such knowledge is typically assumed.

Distributed online knowledge acquisition projects have become quite popular in the past years. Examples include: Freebase,[3] NELL,[4] and ProBase.[5] Other examples include the different projects associated with the Open Mind Initiative, e.g., OMCS, Open Mind Indoor Common Sense [137], which aims to develop intelligent mobile robots for use in home and office environments, and Open Mind Common Sentics [68], a set of GWAPs for the acquisition of affective common-sense knowledge used to enrich SenticNet.

Whereas previous approaches have relied on paid experts or unpaid volunteers, GECKA puts a much stronger emphasis on creating a system that is appealing to a large audience, regardless of whether or not they are interested in contributing to AI. The fundamental aim of GECKA is to transform the activity of entering knowledge into an enjoyable, interactive process as much as possible. Most GWAPs today may be fun to play for a relatively short period of time, but players are not often keen on returning. It goes to say that GWAPs generally evidence a fairly low ‘sticky factor’, defined as the amount of daily active users (DAUs) of an application divided by the number of monthly active users (MAUs).

While MAU on its own is the most-quoted measure of a game’s size, it is only effective in describing size or reach, and not engagement. Similarly, DAU can be a very valuable metric, given that it indicates how much activity a game sees on a daily basis. However, it falls into the same trap as MAU in that it does not discriminate between player-base retention and acquisition. The single-most important metric for engagement is stickiness, i.e., DAU/MAU, which enables more accurate calculation of repeat visits and average knowledge acquired per user (AKAPU).

The key to enhancing a game’s sticky factor, besides great gameplay, is the ability of an application to prompt users to reach out to their friends, e.g., via stories

[3] http://freebase.com

[4] http://rtw.ml.cmu.edu/rtw

[5] http://research.microsoft.com/probase

Fig. 2.6 Outdoor scenario. Game designers can drag&drop objects and characters from the library and specify how these interact with each other (Source: [62])

and pictures about their gameplay. To this end, GECKA allows users to design compelling serious games that can be made available on the App Store for their peers to play (Fig. 2.6). As opposed to traditional GWAPs, GECKA does not limit users to specific-often boring-tasks, but rather gives them the freedom to choose both the kind and the granularity of knowledge to be encoded, through a user-friendly and intuitive interface. This not only improves gameplay and game-stickiness, but also allows common-sense knowledge to be collected in ways that are not predictable a priori.

GECKA is not just a system for the creation of microgames, it is a serious game engine that aims to give designers the means to create long adventure games to be played by others. To this end, GECKA offers functionalities typical of role-play games (RPGs), e.g., a question/answer dialogue box enabling communication and the exchange of objects (optionally tied to correct answers) between players and virtual world inhabitants, a library for enriching scenes with useful and yet visually-appealing objects, backgrounds, characters, and a branching storyline for defining how different game scenes are interconnected.

In the branching story screen, game designers place scene nodes and connect them by defining semantic conditions that specify how the player will move from a scene to another (Fig. 2.7). Making a scene transition may require fulfillment of

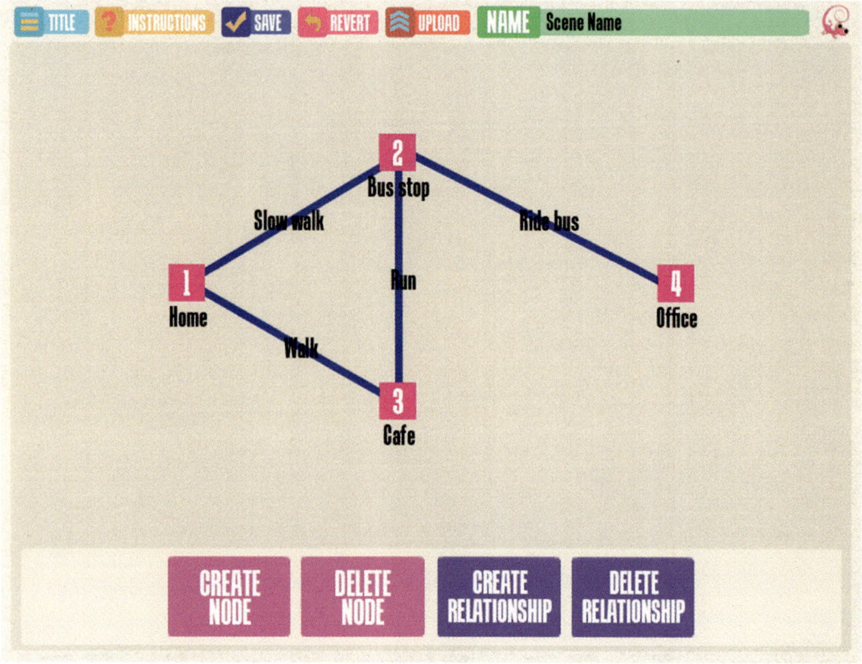

Fig. 2.7 Branching story screen. Game designers can name and connect different scenes according to their semantics and role in the story of the game (Source: [62])

a complex goal, acquisition of an object, or some other relevant condition. These conditions provide invaluable information about the prerequisites of certain actions and the objects that participate in action and goal flows. Goals are created by the combination of smaller semantic primitives ('can', 'cannot', actions, places, and so on), enabling users to specify highly nuanced goals.

Designers can associate goal sequences with each story node through the combination of a set of primitives, actions, objects, and emotions (selected from the library) that describe the end state of the world once the goal sequence is complete. The branching story screen aims to acquire transitional common-sense knowledge, e.g., "if I was at the bus stop before and I am now at the office, I am likely to have taken the bus" and situational common-sense knowledge, e.g., "if you are waiting at the bus stop, your goal is probably to reach a different place".

In case an action or an object are not available in the library, GECKA allows game designers to define their own custom items by building shapes from a set of predefined geometric forms or applying transforms to existing items. This enables the creation of new objects for which there is no available icon by combining available graphics and predefined shapes, and the use of transformations to create various object states, such as a 'broken jar'. The ability of users to create their own custom items and actions is key to maintaining an undisrupted game flow.

Although the aesthetics of a custom object may not be the same as predefined icons, custom objects allow game designers to express their creativity without limiting themselves to the set of available graphics and, hence, allow researchers to discover new common-sense concepts and the semantic features associated with them.

Whenever game designers create a new object or action, they must specify its name and its semantics through prerequisite-outcome-goal (POG) triples, Prerequisites indicate what must be present or have been done before using the object or action. Outcomes include objects or states of the world (including emotional states, e.g., "if I give money to someone, their happiness is likely to rise"). Goals in turn specify the specific scene goals that are facilitated by that particular POG triple. Game designers drag and drop objects and characters from action/object libraries into scenes. For each object, in particular, they can specify a POG triple that describes how such an object is affected by the actions performed over it (Fig. 2.8). POG triples give us pieces of common-sense information like "if I use a can opener on a can, I obtain the content of the can" or "the result of squeezing an orange, is orange juice".

Towards the goal of improving gameplay, and because GECKA mainly aims to collect in typical common-sense knowledge, POG triples associated with a specific object type are shared among all instances of such an object ('inheritance').

Fig. 2.8 Specification of a POG triple. By applying the action 'tie' over a 'pan', in combination with 'stick' and 'lace', a shovel can be obtained (Source: [62])

Table 2.2 List of most common POG triples collected during a pilot testing (Source: [62])

Item	Action	Prerequisite	Outcome	Goal
Orange	Squeeze	–	Orange juice	Quench thirst
Bread	Cut	Knife	Bread slices	–
Bread slices	Stack	Ham, mayonnaise	Sandwich	Satisfy hunger
Coffee beans	Hit	Pestle	Coffee powder	–
Coffee maker	Fill	Coffee powder, water	Coffee	–
Bottle	Fill	Water	Bottled water	Quench thirst
Chair	Hit	Hammer	Wood pieces	–
Can	Open	Can opener	Food	Satisfy hunger
Towel	Cut	Scissors	Bandage	–
Sack	Fill	Sand	Sandbag	Flood control

Whenever a game designer associates a POG to an object in the scene, that POG instantly becomes shared among all the other objects of the same type, no matter if these are located in different scenes. New instances inherit this POG as well.

Game designers, however, can create exceptions of any object type through the creation of new custom objects. A `moldy_bread` custom object, for example, normally inherits all the POGs of `bread` but these can be changed, modified, or removed at the time of object instantiation without affecting other `bread` type objects. The POG specification is one of the most effective means for collecting common-sense knowledge, given that it is performed quite often by the game designer during the creation of scenes (Fig. 2.9).

From a simple POG definition we may obtain a large amount of knowledge, including interaction semantics between different objects, prerequisites of actions, and the goals commonly associated with such actions (Table 2.2). These pieces of common-sense knowledge, are very clearly-structured, and thus easy to assimilate into the knowledge base, due to the fixed framework for defining interaction semantics. POG specifications not only allow game designers to define interaction semantics between objects but also to specify how the original player, action/object recipients, and non-recipients react to various actions by setting parameters involving character health, hunger, pleasantness, and sensitivity (Fig. 2.10). While the first two parameters allow more physiological common-sense knowledge to be collected, pleasantness and sensitivity directly map affective common-sense knowledge onto the Hourglass model. This is, in turn, used to enhance reasoning within SenticNet, especially for tasks such as emotion recognition, goal inference, and sentiment analysis.

2.2 Knowledge Representation

This section describes how the knowledge collected from OMCS, WNA, and GECKA is represented redundantly at three levels: semantic network, matrix, and vector space. In particular, the collected or crowd sourced pieces of knowledge

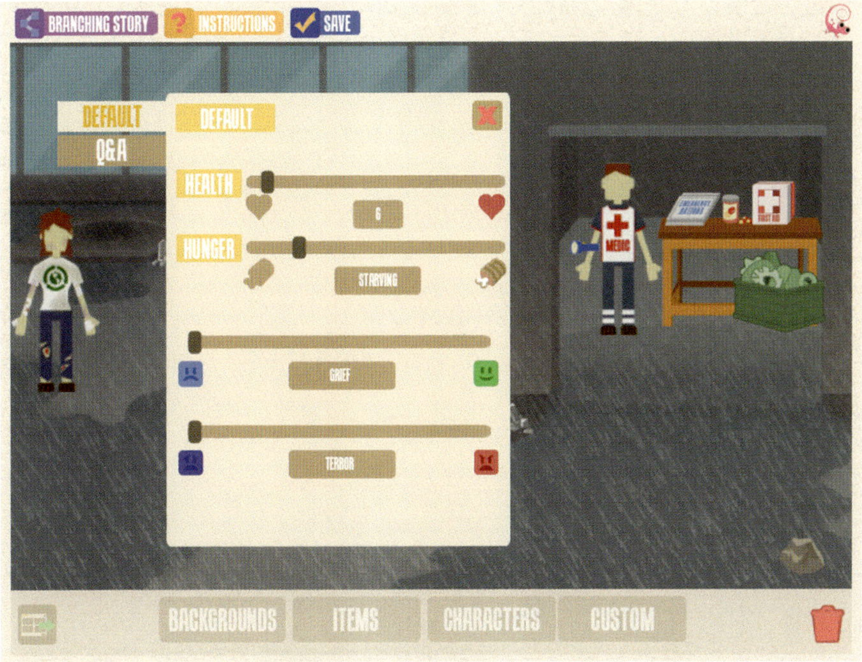

Fig. 2.9 Status of a new character in the scene who is ill and extremely hungry, plus has very low levels of pleasantness (grief) and sensitivity (terror) (Source: [62])

are firstly integrated in a semantic network as triples of the format <concept-relationship-concept> (Sect. 2.2.1). Secondly, the graph is represented as a matrix having concepts as rows and the combination <relationship-concept> as columns (Sect. 2.2.1). Finally, multi-dimensionality reduction is applied to such a matrix in order to create a vector space representation of common-sense knowledge (Sect. 2.2.3).

2.2.1 AffectNet Graph

AffectNet (Fig. 2.11) is an affective common-sense knowledge base mainly built upon ConceptNet [195], the graph representation of the Open Mind corpus, which is structurally similar to WordNet, but whose scope of contents is general world knowledge, in the same vein as Cyc. Instead of insisting on formalizing common-sense reasoning using mathematical logic [220], ConceptNet uses a new approach: it represents data in the form of a semantic network and makes it available for use in natural language processing. The prerogative of ConceptNet, in fact, is contextual common-sense reasoning: while WordNet is optimized for lexical categorization

```xml
<scenes>
    <sceneData>
        <sceneType>
            <string>kitchen</string>
        </sceneType>
        <items>
            <itemData>
                <itemType>
                    <string>bread slices</string>
                </itemType>
                <position>
                    <x>8.04757</x>
                    <y>2.32971239</y>
                </position>
                <actions>
                    <actionData>
                        <actionType>
                            <string>stack</string>
                        </actionType>
                        <POG_Data>
                            <prerequisites>
                                <string>ham</string>
                                <string>mayonnaise</string>
                            </prerequisites>
                            <outcomes>
                                <string>sandwich</string>
                            </outcomes>
                            <goal>
                                <string>satisfy hunger</string>
                            </goal>
                        </POG_Data>
                        <player>
                            <affect>
                                <health>80</health>
                                <hunger>50</hunger>
                                <pleasantness>5</pleasantness>
                                <sensitivity>3</sensitivity>
                            </affect>
                        </player>
                        <recipientCharacter>
                            <type>hungry man</type>
                            <affect>
                                <health>80</health>
                                <hunger>50</hunger>
                                <pleasantness>5</pleasantness>
                                <sensitivity>3</sensitivity>
                            </affect>
                        </recipientCharacter>
                        <nonRecipientCharacter>
```


Fig. 2.10 A sample XML output deriving from the creation of a scene in GECKA. Actions are collected and encoded according to their semantics (Source: [62])

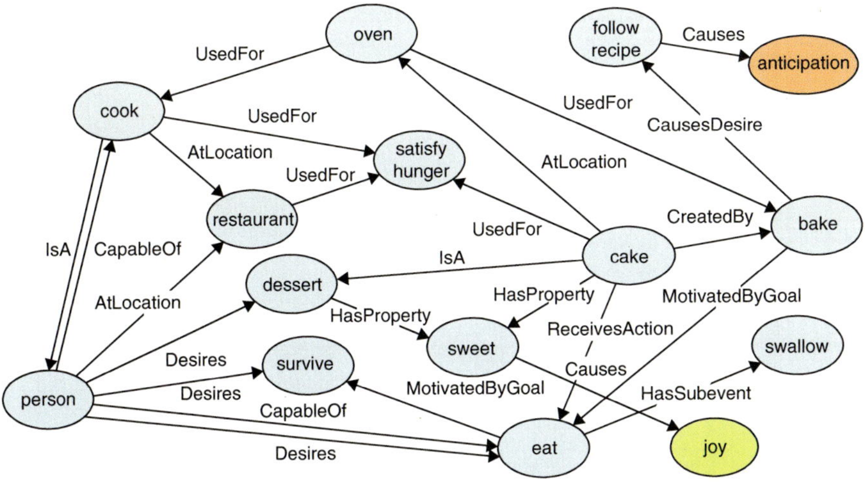

Fig. 2.11 A sketch of the AffectNet graph showing part of the semantic network for the concept `cake`. The directed graph not only specifies semantic relations between concepts but also connects these to affective nodes (Source: The Authors)

and word-similarity determination, and Cyc is optimized for formalized logical reasoning, ConceptNet is optimized for making practical context-based inferences over real-world texts.

In ConceptNet, WordNet's notion of a node in the semantic network is extended from purely lexical items (words and simple phrases with atomic meaning) to include higher-order compound concepts, e.g., '`satisfy hunger`' and '`follow recipe`', to represent knowledge around a greater range of concepts found in everyday life (Table 2.3). Moreover WordNet's repertoire of semantic relations is extended from the triplet of synonym, *IsA* and *PartOf*, to a repertoire of 20 semantic relations including, for example, *EffectOf* (causality), *SubeventOf* (event hierarchy), *CapableOf* (agent's ability), *MotivationOf* (affect), *PropertyOf*, and *LocationOf*. ConceptNet's knowledge is also of a more informal, defeasible, and practically valued nature.

For example, WordNet has formal taxonomic knowledge that 'dog' is a 'canine', which is a 'carnivore', which is a 'placental mammal'; but it cannot make the practically oriented member-to-set association that 'dog' is a 'pet'. ConceptNet also contains a lot of knowledge that is defeasible, i.e., it describes something that is often true but not always, e.g., EffectOf('fall off bicycle', 'get hurt'), which is something we cannot leave aside in common-sense reasoning. Most of the facts interrelating ConceptNet's semantic network are dedicated to making rather generic connections between concepts. This type of knowledge can be brought back to Minsky's K-lines, as it increases the connectivity of the semantic network and makes it more likely that concepts parsed out of a text document can be mapped into ConceptNet.

Table 2.3 Comparison between WordNet and ConceptNet. While WordNet synsets contain vocabulary knowledge, ConceptNet assertions convey knowledge about what concepts are used for (Source: [50])

Term	WordNet Hypernyms	ConceptNet Assertions
Cat	Feline; Felid; Adult male; Man; Gossip; Gossiper; Gossipmonger; Rumormonger; Rumourmonger; Newsmonger; Woman; Adult female; Stimulant; Stimulant drug; Excitant; Tracked vehicle; ...	Cats can hunt mice; Cats have whiskers; Cats can eat mice; Cats have fur; cats have claws; Cats can eat meat; cats are cute; ...
Dog	Canine; Canid; Disagreeable woman; Chap; Fellow; Feller; Lad; Gent; Fella; Scoundrel; Sausage; Follow, ...	Dogs are mammals; A dog can be a pet; A dog can guard a house; You are likely to find a dog in kennel; An activity a dog can do is run; A dog is a loyal friend; A dog has fur; ...
Language	Communication; Auditory communication; Word; Higher cognitive process; Faculty; Mental faculty; Module; Text; Textual matter; ...	English is a language; French is a language; Language is used for communication; Music is a language; A word is part of language; ...
iPhone	N/A;	An iPhone is a kind of telephone; An iPhone is a kind of computer; An IPhone can display your position on a map; An IPhone can send and receive emails; An IPhone can display the time; ...
Birthday gift	Present;	Card is birthday gift; Present is birthday gift; Buying something for a loved one is for a birthday gift; ...

ConceptNet is produced by an automatic process, which first applies a set of extraction rules to the semi-structured English sentences of the OMCS corpus, and then applies an additional set of 'relaxation' procedures, i.e., filling in and smoothing over network gaps, to optimize the connectivity of the semantic network. In ConceptNet 2, a new system for weighting knowledge was implemented, which scores each binary assertion based on how many times it was uttered in the OMCS corpus, and on how well it can be inferred indirectly from other facts in ConceptNet. In ConceptNet 3 [142], users can also participate in the process of refining knowledge by evaluating existing statements on Open Mind Commons [296], the new interface for collecting common-sense knowledge from users over the Web.

By giving the user many forms of feedback and using inferences by analogy to find appropriate questions to ask, Open Mind Commons can learn well-connected structures of common-sense knowledge, refine its existing knowledge, and build analogies that lead to even more powerful inferences. ConceptNet 4 includes data that was imported from the online game Verbosity. It also includes the initial import of the Chinese ConceptNet. ConceptNet 5 [297], eventually, contains knowledge from English Wikipedia, specifically from DBPedia, which extracts knowledge from the info-boxes that appear on articles, and ReVerb, a machine-reading project extracting relational knowledge from the actual text of each article. It also includes a large amount of content from the English Wiktionary, including synonyms, antonyms, translations of concepts into hundreds of languages, and multiple labeled word senses for many English words.

ConceptNet 5 contains more dictionary-style knowledge coming from WordNet and some knowledge about people's intuitive word associations coming from GWAPs. Previous versions of ConceptNet have been distributed as idiosyncratic database structures plus some software to interact with them. ConceptNet 5 is not a piece of software or a database, but a hypergraph, that is, a graph that has edges about edges. Each statement in ConceptNet, in fact, has justifications pointing to it, explaining where it comes from and how reliable the information seems to be. ConceptNet is a good source of common-sense knowledge but alone is not enough for sentiment analysis tasks as it specifies how concepts are semantically related to each other but often lacks connections between concepts that convey the same kind of emotion or similar polarity. To overcome such a hurdle, affective knowledge from WNA is added.

2.2.2 AffectNet Matrix

In Chinese culture (and many others), the concepts of 'heart' and 'mind' used to be expressed by the same word (心) as it was believed that consciousness and thought came from the cardiac muscle. In human cognition, in fact, thinking and feeling are mutually present: emotions are often the product of our thoughts, as well as our reflections are often the product of our affective states. Emotions are intrinsically part of our mental activity and play a key role in communication and decision-making processes. Emotion is a chain of events made up of feedback loops. Feelings and behavior can affect cognition, just as cognition can influence feeling. Emotion, cognition, and action interact in feedback loops and emotion can be viewed in a structural model tied to adaptation [246].

There is actually no fundamental opposition between emotion and reason. In fact, it may be argued that reason consists of basing choices on the perspectives of emotions at some later time. Reason dictates not giving in to one's impulses because doing so may cause greater suffering later [131]. Reason does not necessarily imply exertion of the voluntary capacities to suppress emotion. It does not necessarily involve depriving certain aspects of reality of their emotive powers.

On the contrary, our voluntary capacities allow us to draw more of reality into the sphere of emotion. They allow one's emotions to be elicited not merely by the proximal, or the perceptual, or that which directly interferes with one's actions, but by that which, in fact, touches on one's concerns, whether proximal or distal, whether occurring now or in the future, whether interfering with one's own life or that of others. Cognitive functions serve emotions and biological needs. Information from the environment is evaluated in terms of its ability to satisfy or frustrate needs. What is particularly significant is that each new cognitive experience that is biologically important is connected with an emotional reaction such as fear, pleasure, pain, disgust, or depression [226].

In order to build a semantic network that contains both semantic and affective knowledge, ConceptNet and WNA are blended together by combining the matrix representations of the two knowledge bases linearly into a single matrix, in which the information between the two initial sources is shared. The first step to create the affective blend is to transform the input data so that it can all be represented in the same matrix. To do this, the lemma forms of ConceptNet concepts are aligned with the lemma forms of the words in WNA and the most common relations in the affective knowledge base are mapped into ConceptNet's set of relations, e.g., Hypernym into *IsA* and Holonym into *PartOf*. In particular, ConceptNet is first converted into a matrix by dividing each assertion into two parts: a concept and a feature, where a feature is simply the assertion with the first or the second concept left unspecified such as 'a wheel is part of' or 'is a kind of liquid'. The entries in the resulting matrix are positive or negative numbers, depending on the reliability of the assertions, and their magnitude increases logarithmically with the confidence score. WNA, similarly, is represented as a matrix where rows are affective concepts and columns are features related to these.

The result of aligning the matrix representations of ConceptNet and WNA is a new affective semantic network, in which common-sense concepts are linked to a hierarchy of affective domain labels. In such a semantic network, termed AffectNet,[6] common-sense and affective knowledge are in fact combined, not just concomitant, i.e., everyday life concepts like `have_breakfast`, `meet_people`, or `watch_tv` are linked to affective domain labels like 'joy', 'anger', or 'surprise'.

Because the AffectNet graph is made of triples of the format <concept-relationship-concept>, the entire knowledge repository can be visualized as a large matrix, with every known concept of some statement being a row and every known semantic feature (relationship+concept) being a column. Such a representation has several advantages including the possibility to perform cumulative analogy [81, 310]. Cumulative analogy is performed by first selecting a set of nearest neighbors, in terms of similarity, of the input concept and then by projecting known properties of this set onto not known properties of the concept (Table 2.4).

It is inherent to human nature to try to categorize things and people, finding patterns and forms they have in common. One of the most intuitive ways to relate two entities is through their similarity. Similarity is one of the six Gestalt principles which guide the human perception of the world, the remaining ones being: Proximity, Closure, Good Continuation, Common Fate, and Good Form. According to Merriam Webster, 'similarity' is a quality that makes one person or thing like another and 'similar' means having characteristics in common. There are many ways in which objects can be perceived as similar, such as having similar

[6]http://sentic.net/affectnet.zip

Table 2.4 Cumulative analogy allows for the inference of new pieces of knowledge by comparing similar concepts. In the example, it is inferred that the concept `special_occasion` causes joy as it shares the same set of semantic features with `wedding` and `birthday` (which also cause joy) (Source: The Authors)

	Semantic Features (*relationship+concept*)					
Concepts	...	*Causes* joy	*IsA* event	*UsedFor* housework	*MotivatedBy* celebration	...
⋮		⋮	⋮	⋮	⋮	
wedding	...	x	x	–	x	...
broom	...	–	–	x	–	...
special_occasion	...	**x?**	**x**	**–**	**x**	...
birthday	...	x	x	–	x	...
⋮		⋮	⋮	⋮	⋮	

color, shape, size, and texture. If we move away from mere visual stimuli, we can apply the same principles to define a similarity between concepts based on shared semantic features.

For AffectNet, however, such a process is rather time- and resource-consuming as its matrix representation is made of several thousands columns (fat matrix). In order to perform analogical reasoning in a faster and more efficient manner, such a matrix can be represented as a vector space by applying multi-dimensionality reduction techniques that decrease the number of semantic features associated with each concept without compromising too much knowledge representation.

2.2.3 *AffectiveSpace*

The best way to solve a problem is to know an *a-priori* solution for it. But, if we have to face a problem we have never encountered before, we need to use our intuition. Intuition can be explained as the process of making analogies between the current problem and the ones solved in the past to find a suitable solution. Marvin Minsky attributes this property to the so called 'difference-engines' [212]. This particular kind of agent operates by recognizing differences between the current state and the desired state, and acts to reduce each difference by invoking K-lines that turn on suitable solution methods. This kind of thinking is maybe the essence of our supreme intelligence since in everyday life no two situations are ever the same and have to perform this action continuously.

To emulate such a process, AffectiveSpace[7] [44], a novel affective common-sense knowledge visualization and analysis system, is used. The human mind constructs intelligible meanings by continuously compressing over vital relations [124]. The compression principles aim to transform diffuse and distended conceptual structures to more focused versions so as to become more congenial for human understanding. To this end, principal component analysis (PCA) has been applied on the matrix representation of AffectNet. In particular, truncated singular value decomposition (TSVD) has been preferred to other dimensionality reduction techniques for its simplicity, relatively low computational cost, and compactness.

TSVD, in fact, is particularly suitable for measuring the cross-correlations between affective common-sense concepts as it uses an orthogonal transformation to convert the set of possibly correlated common-sense features associated with each concept into a set of values of uncorrelated variables (the principal components of the SVD). By using Lanczos' method [176], moreover, the generalization process is relatively fast (a few seconds), despite the size and the sparseness of AffectNet. The objective of such compression is to allow many details in the blend of ConceptNet and WNA to be removed such that the blend only consists of a few essential features that represent the global picture. Applying TSVD on AffectNet, in fact, causes it to describe other features that could apply to known affective concepts by analogy: if a concept in the matrix has no value specified for a feature owned by many similar concepts, then by analogy the concept is likely to have that feature as well. In other words, concepts and features that point in similar directions and, therefore, have high dot products, are good candidates for analogies. A pioneering work on understanding and visualizing the affective information associated with natural language text was conducted by Osgood et al. [231]. Osgood used multi-dimensional scaling (MDS) to create visualizations of affective words based on similarity ratings of the words provided to subjects from different cultures. Words can be thought of as points in a multi-dimensional space and the similarity ratings represent the distances between these words. MDS projects these distances to points in a smaller dimensional space (usually two or three dimensions). Similarly, AffectiveSpace aims to grasp the semantic and affective similarity between different concepts by plotting them into a multi-dimensional vector space [55]. Unlike Osgood's space, however, the building blocks of AffectiveSpace are not simply a limited set of similarity ratings between affect words, but rather millions of confidence scores related to pieces of common-sense knowledge linked to a hierarchy of affective domain labels. Rather than merely determined by a few human annotators and represented as a word-word matrix, in fact, AffectiveSpace is built upon an affective common-sense knowledge base, namely AffectNet, represented as a concept-feature matrix. After performing TSVD on such matrix, herein after termed A for the sake of conciseness, a low-rank approximation of it is obtained, that is, a new matrix $\tilde{A} = U_k \Sigma_k V_k^T$. This approximation is based on minimizing the Frobenius norm of the difference between A and $\tilde{A}$ under the constraint $rank(\tilde{A}) = k$. For the Eckart–

[7] http://sentic.net/affectivespace.zip

Young theorem [113], it represents the best approximation of A in the least-square sense, in fact:

$$\min_{\tilde{A}|rank(\tilde{A})=k} |A - \tilde{A}| = \min_{\tilde{A}|rank(\tilde{A})=k} |\Sigma - U^*\tilde{A}V| = \min_{\tilde{A}|rank(\tilde{A})=k} |\Sigma - S| \tag{2.1}$$

assuming that $\tilde{A}$ has the form $\tilde{A} = USV^*$, where S is diagonal. From the rank constraint, i.e., S has k non-zero diagonal entries, the minimum of the above statement is obtained as follows:

$$\min_{\tilde{A}|rank(\tilde{A})=k} |\Sigma - S| = \min_{s_i} \sqrt{\sum_{i=1}^{n} (\sigma_i - s_i)^2} \tag{2.2}$$

$$\min_{s_i} \sqrt{\sum_{i=1}^{n} (\sigma_i - s_i)^2} = \min_{s_i} \sqrt{\sum_{i=1}^{k} (\sigma_i - s_i)^2 + \sum_{i=k+1}^{n} \sigma_i^2} = \sqrt{\sum_{i=k+1}^{n} \sigma_i^2} \tag{2.3}$$

Therefore, $\tilde{A}$ of rank k is the best approximation of A in the Frobenius norm sense when $\sigma_i = s_i$ ($i = 1, \ldots, k$) and the corresponding singular vectors are the same as those of A. If all but the first k principal components are discarded, common-sense concepts and emotions are represented by vectors of k coordinates. These coordinates can be seen as describing concepts in terms of 'eigenmoods' that form the axes of AffectiveSpace, i.e., the basis $e_0, \ldots, e_{k-1}$ of the vector space (Fig. 2.12). For example, the most significant eigenmood, e_0, represents concepts with positive affective valence. That is, the larger a concept's component in the e_0 direction is, the more affectively positive it is likely to be. Concepts with negative e_0 components, then, are likely to have negative affective valence. Thus, by exploiting the information sharing property of TSVD, concepts with the same affective valence are likely to have similar features – that is, concepts conveying the same emotion tend to fall near each other in AffectiveSpace.

Concept similarity does not depend on their absolute positions in the vector space, but rather on the angle they make with the origin. For example concepts such as `beautiful_day`, `birthday_party`, and `make_person_happy` are found very close in direction in the vector space, while concepts like `feel_guilty`, `be_laid_off`, and `shed_tear` are found in a completely different direction (nearly opposite with respect to the centre of the space).

The problem with this kind of representation is that it is not scalable: when the number of concepts and of semantic features grows, the AffectNet matrix becomes too high-dimensional and too sparse for SVD to be computed [21]. Although there has been a body of research on seeking for fast approximations of the SVD, the approximate methods are at most $\approx$ 5 times faster than the standard one [210], making it unattractive for real-world big data applications.

It has been conjectured that there might be simple but powerful meta-algorithms underlying neuronal learning [184]. These meta-algorithms should be fast, scalable,

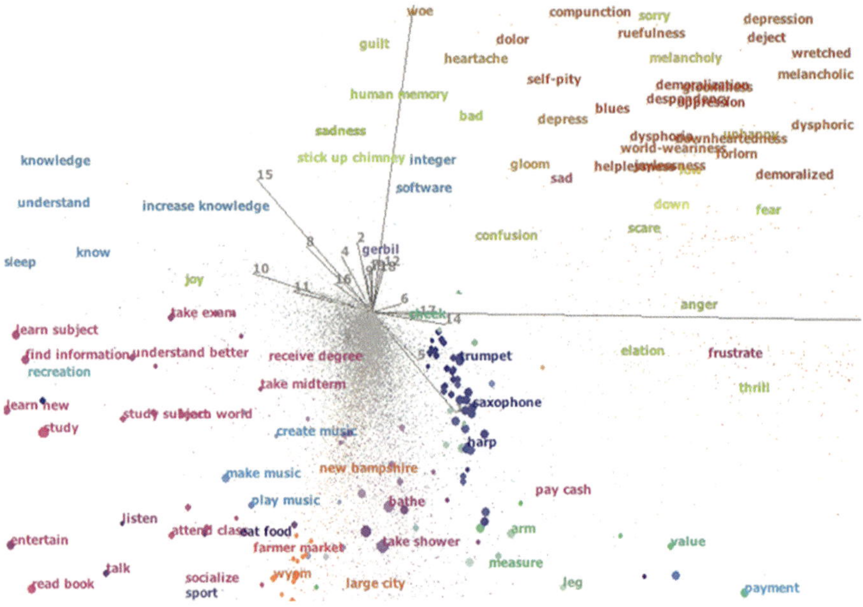

Fig. 2.12 A sketch of AffectiveSpace. Affectively positive concepts (in the bottom-left corner) and affectively negative concepts (in the up-right corner) are floating in the multi-dimensional vector space (Source: [44])

effective, with few-to-no specific assumptions, and biologically plausible [21]. Optimizing all the $\approx 10^{15}$connections through the last few million years' evolution is very unlikely [21]. Alternatively, nature probably only optimizes the global connectivity (mainly the white matter), but leaves the other details to randomness [21]. In order to cope with the ever-growing number of concepts and semantic features, thus, SVD is replaced with random projection (RP) [29], a data-oblivious method, to map the original high-dimensional data-set into a much lower-dimensional subspace by using a Gaussian N(0, 1) matrix, while preserving the pair-wise distances with high probability. This theoretically solid and empirically verified statement follows Johnson Lindenstrauss (JL) Lemma [21]. The JL Lemma states that with high probability, for all pairs of points x, y $\in$ X simultaneously

$$\sqrt{\frac{m}{d}} \parallel x - y \parallel_2 (1 - \varepsilon) \leq \parallel \Phi x - \Phi y \parallel_2 \leq \tag{2.4}$$

$$\leq \sqrt{\frac{m}{d}} \parallel x - y \parallel_2 (1 + \varepsilon), \tag{2.5}$$

where X is a set of vectors in Euclidean space, d is the original dimension of this Euclidean space, m is the dimension of the space we wish to reduce the data points to, ε is a tolerance parameter measuring to what extent is the maximum allowed distortion rate of the metric space, and Φ is a random matrix.

Structured random projection for making matrix multiplication much faster was introduced in [279]. Achlioptas [2] proposed *sparse random projection* to replace the Gaussian matrix with i.i.d. entries in

$$\phi_{ji} = \sqrt{s}\begin{cases} 1 & \text{with prob. } \frac{1}{2s} \\ 0 & \text{with prob.} 1 - \frac{1}{s} \\ -1 & \text{with prob.} \frac{1}{2s} \end{cases}, \tag{2.6}$$

where one can achieve a ×3 speedup by setting $s = 3$, since only $\frac{1}{3}$ of the data need to be processed. However, since AffectNet is already too sparse, using sparse random projection is not advisable.

When the number of features is much larger than the number of training samples ($d \gg n$), subsampled randomized Hadamard transform (SRHT) is preferred, as it behaves very much like Gaussian random matrices but accelerates the process from $\mathcal{O}(n\,d)$ to $\mathcal{O}(n\,log\,d)$ time [197]. Following [197, 309], for $d = 2^p$ where p is any positive integer, a SRHT can be defined as:

$$\Phi = \sqrt{\frac{d}{m}}\text{RHD} \tag{2.7}$$

where

- m is the number we want to subsample from d features randomly.
- R is a random $m \times d$ matrix. The rows of R are m uniform samples (without replacement) from the standard basis of $\mathbb{R}^d$.
- H∈ $\mathbb{R}^{dxd}$ is a normalized Walsh-Hadamard matrix, which is defined recursively: $H_d = \begin{bmatrix} H_{d/2} & H_{d/2} \\ H_{d/2} & H_{d/2} \end{bmatrix}$ with $H_2 = \begin{bmatrix} +1 & +1 \\ +1 & -1 \end{bmatrix}$.
- D is a $d \times d$ diagonal matrix and the diagonal elements are i.i.d. Rademacher random variables.

The subsequent analysis only relies on the distances and angles between pairs of vectors (i.e. the Euclidean geometry information), and it is sufficient to set the projected space to be logarithmic in the size of the data [10] and apply SRHT.

The key to performing common-sense reasoning is to find a good trade-off for representing knowledge. Since, in life, two situations are never exactly the same, no representation should be too concrete, or it will not apply to new situations, but, at the same time, no representation should be too abstract, or it will suppress too many details. AffectNet already supports different representations, in fact, it maintains different ways of conveying the same idea with redundant concepts, e.g., `car` and `automobile`, that can be reconciled through background linguistic knowledge, if necessary. Within AffectiveSpace, this knowledge representation trade-off can be seen in the choice of the vector space dimensionality.

The number k of singular values selected to build AffectiveSpace, in fact, is a measure of the trade-off between precision and efficiency in the representation of the affective common-sense knowledge base. The bigger k is, the more precisely AffectiveSpace represents AffectNet's knowledge, but generating the vector space is slower, as is computing of dot products between concepts. The smaller k is, on the other hand, the more efficiently AffectiveSpace represents affective common-sense knowledge both in terms of vector space generation and of dot product computation. However, too few dimensions risk not to correctly represent AffectNet as concepts defined with too few features tend to be too close to each other in the vector space and, hence, not easily distinguishable and clusterable. In order to find a good k, AffectiveSpace was tested on a benchmark for affective common-sense knowledge (BACK) built by applying CF-IOF (concept frequency – inverse opinion frequency) [51] on the 5,000 posts of the LiveJournal corpus (Table 2.5).

CF-IOF is a technique that identifies common domain-dependent semantics in order to evaluate how important a concept is to a set of opinions concerning the same topic. Firstly, the frequency of a concept c for a given domain d is calculated by counting the occurrences of the concept c in the set of available d-tagged opinions and dividing the result by the sum of number of occurrences of all concepts in the set of opinions concerning d. This frequency is then multiplied by the logarithm of the inverse frequency of the concept in the whole collection of opinions, that is:

$$CF\text{-}IOF_{c,d} = \frac{n_{c,d}}{\sum_k n_{k,d}} \log \sum_k \frac{n_k}{n_c} \tag{2.8}$$

Table 2.5 Some examples of LiveJournal posts where affective information is not conveyed explicitly through affect words (Source: [50])

Mood	LiveJournal Posts	Concepts
Happy	Finally I got my student cap! I am officially high school graduate now! Our dog Tanja, me, Timo (our art teacher) and EmmaMe, Tanja, Emma and Tiia Only two weeks to Japan!!	Student; school graduate; Japan
Happy	I got a kitten as an early birthday gift on Monday. Abby was smelly, dirty, and gnawing on the metal bars of the kitten carrier though somewhat calm when I picked her up. We took her. She threw up on me on the ride home and repeatly keeps sneesing in my face.	Kitten; birthday gift; metal bar; face
Sad	Hi. Can I ask a favor from you? This will only take a minute. Please pray for Marie, my friends' dog a labrador, for she has canine distemper. Her lower half is paralysed and she's having locked jaw. My friends' family is feeding her through syringe.	Friends; dog; labrador; canine distemper; jaw; syringe
Sad	My uncle paul passed away on febuary 16, 2008. he lost his battle with cancer. i remember spending time with him and my aunt nina when they babysat me. we would go to taco bell to eat nachos.	Uncle; battle; cancer; aunt; taco bell; nachos

Table 2.6 Distribution of concepts through the Pleasantness dimension. The affective information associated with most concepts concentrates around the centre of the Hourglass, rather than its extremes (Source: [50])

Level	Label	Frequency (%)
G(−1)	Grief	14.3
G(−2/3)	Sadness	19.8
G(−1/3)	Pensiveness	11.4
0	Neutral	10.5
G(1/3)	Serenity	20.6
G(2/3)	Joy	18.3
G(1)	Ecstasy	5.1

where $n_{c,d}$ is the number of occurrences of concept c in the set of opinions tagged as d, n_k is the total number of concept occurrences, and n_c is the number of occurrences of c in the whole set of opinions. A high weight in CF-IOF is reached by a high concept frequency in a given domain and a low frequency of the concept in the whole collection of opinions. Specifically, CF-IOF weighting was exploited to filter out common concepts in the LiveJournal corpus and to detect relevant mood-dependent semantics for the set of 24 emotions defined by Plutchik [246]. The result was a benchmark of 2000 affective concepts that were screened by 21 English-speaking students who were asked to map each concept to the 24 different emotional categories, which form the Hourglass of Emotions [57] (explained later). Results obtained were averaged (Table 2.6).

BACK's concepts were compared with the classification results obtained by applying the AffectiveSpace process using different values of k, from 1 to 250. As shown in Fig. 2.13, the best trade-off is achieved at 100, as selecting more than 100 singular values does not improve accuracy significantly.

The distribution of the values of each AffectiveSpace dimension is bell-shaped, with different centers and degrees of dispersion around them. Affective common-sense concepts, in fact, tend to be close to the origin of the vector space (Fig. 2.14). In order to more uniformly distribute concept density in AffectiveSpace, an alternative strategy to represent the vector space was investigated. Such strategy consists in centring the values of the distribution of each dimension on the origin and in mapping dimensions according to a transformation $x \in \mathbb{R} \mapsto x^* \in [-1, 1]$.

This transformation is often pivotal for better clustering AffectiveSpace as the vector space tends to have different grades of dispersion of data points across different dimensions, with some space regions more densely populated than others. The switch to a different space configuration helps to distribute data more uniformly, possibly leading to an improved (or, at least, different) reasoning process. In particular, the transformation $x_{ij} \mapsto x_{ij} - \mu_i$ is first applied, being μ_i the average of all values of the i-th dimension. Then a normalization is applied, combining the previous transformation with a new one $x_{ij} \mapsto \frac{x_{ij}}{a \cdot \sigma_i}$, where σ_i is the standard deviation calculated on the i-th dimension and a is a coefficient that can modify the same proportion of data that is represented within a specified interval.

Finally, in order to ensure that all components of the vectors in the defined space are within $[-1, 1]$ (i.e., that the Chebyshev distance between the origin and each

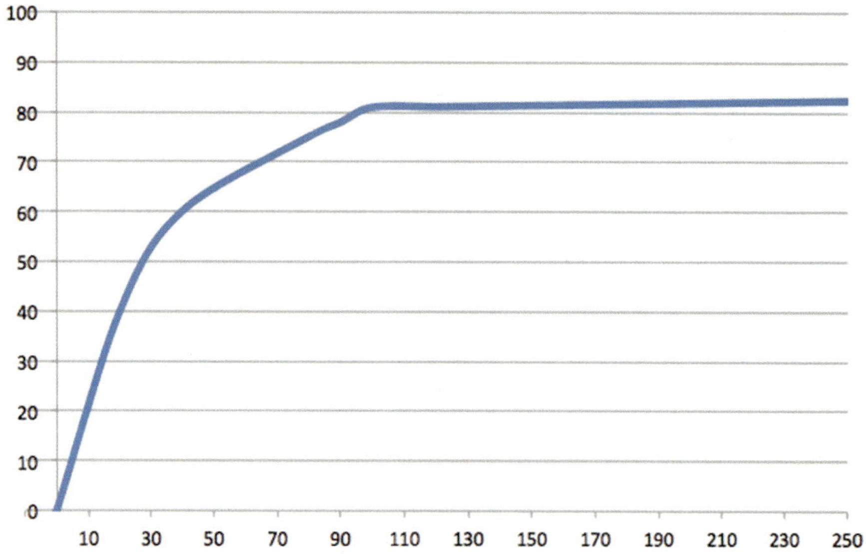

Fig. 2.13 Accuracy values achieved by testing AffectiveSpace on BACK, with dimensionality spanning from 1 to 250. The best trade-off between precision and efficiency is obtained around 100 (Source: [50])

vector is smaller or equal to 1), a final transformation $x_{ij} \mapsto s(x_{ij})$ is needed, where $s(x)$ is a sigmoid function. Different choices for the sigmoid function may be made, influencing how 'fast' the function approaches the unit value while the independent variable approaches infinity. Combining the proposed transformations, two possible mapping functions are expressed in the following formulae 2.9 and 2.10:

$$x^*_{ij} = tanh\left(\frac{x_{ij} - \mu_i}{a \cdot \sigma_i}\right) \tag{2.9}$$

$$x^*_{ij} = \frac{x_{ij} - \mu_i}{a \cdot \sigma_i + \left|x_{ij} - \mu_i\right|} \tag{2.10}$$

This space transformation leads to two main advantages, which could be of notable importance depending on the problem being tackled. Firstly, this different space configuration ensures that each dimension is equally important by avoiding that the information provided by dimensions with higher (i.e., more distant from the origin) averages predominates. Secondly, normalizing according to the standard deviations of each dimension allows for a more uniform distribution of data around the origin, leading to a full use of information potential.

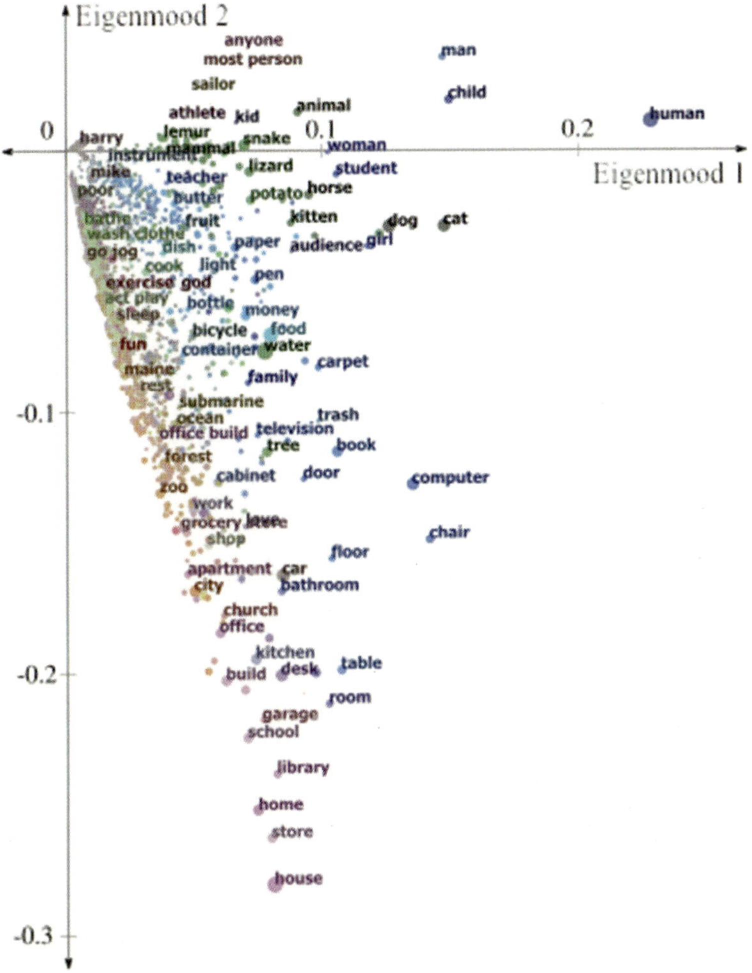

Fig. 2.14 A two-dimensional projection (first and second eigenmoods) of AffectiveSpace. From this visualization, it is evident that concept density is usually higher near the centre of the space (Source: [50])

2.3 Knowledge-Based Reasoning

This section describes the techniques adopted for generating semantics and sentics from the three different common-sense knowledge representations described above. In particular, semantics are inferred by means of spreading activation (Sect. 2.3.1) while sentics are created through the ensemble application of an emotion categorization model (Sect. 2.3.2) and a set of neural networks (Sect. 2.3.3).

2.3.1 *Sentic Activation*

An important difference between traditional AI systems and human intelligence is our ability to harness common sense knowledge gleaned from a lifetime of learning and experiences to inform our decision-making and behavior. This allows humans to adapt easily to novel situations where AI fails catastrophically for lack of situation-specific rules and generalization capabilities. In order for machines to exploit common sense knowledge in reasoning as humans do, moreover, we need to endow them with human-like reasoning strategies. In problem-solving situations, in particular, several analogous representations of the same problem should be maintained in parallel while trying to solve it so that, when problem-solving begins to fail while using one representation, the system can switch to one of the others [60].

Sentic activation [59] is a two-level reasoning framework for the generation of semantics (Fig. 2.15). By representing common-sense knowledge redundantly at three levels (semantic network, matrix, and vector space), sentic activation implements a reasoning loop that solves the problem of relevance in spreading activation by guiding the activation of nodes through analogical reasoning. In

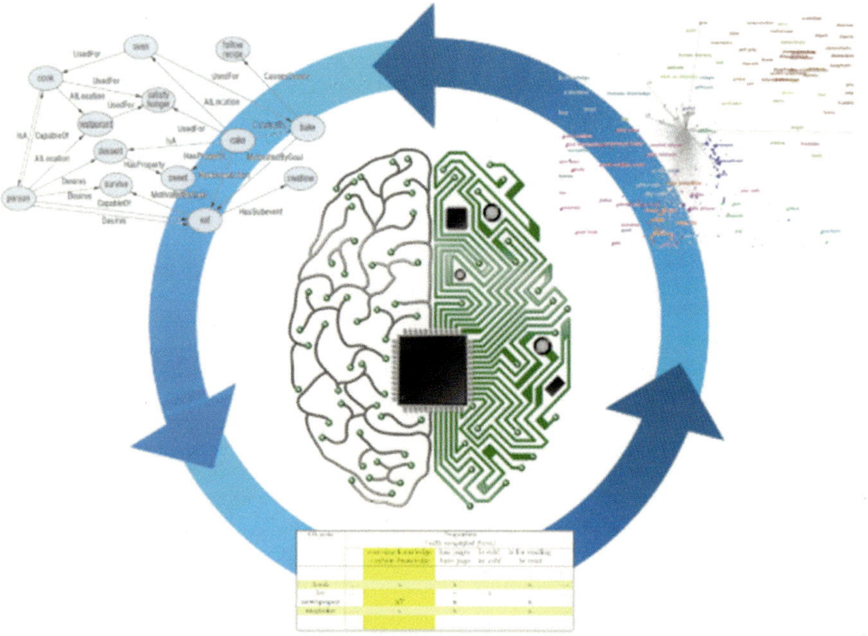

Fig. 2.15 The sentic activation loop. Common-sense knowledge is represented redundantly at three levels (semantic network, matrix, and vector space) in order to solve the problem of relevance in spreading activation (Source: The Authors)

particular, the framework limits the activation of concepts in AffectNet by exploiting the set of semantically related concepts generated by AffectiveSpace.

Sentic activation is inspired by the current thinking in cognitive psychology, which suggests that humans process information at a minimum of two distinct levels. There is extensive evidence for the existence of two (or more) processing systems within the human brain, one that involves fast, parallel, unconscious processing, and one that involves slow, serial, more conscious processing [71, 119, 170, 289]. Dual-process models of automatic and controlled social cognition have been proposed in nearly every domain of social psychology. Evidence from neurosciences supports this separation, with identifiably different brain regions involved in each of the two systems [192].

Such systems, termed U-level (unconscious) and C-level (conscious), can operate simultaneously or sequentially, and are most effective in different contexts. The former, in particular, works intuitively, effortlessly, globally, and emotionally (Sect. 2.3.1.1). The latter, in turn, works logically, systematically, effortfully, and rationally (Sect. 2.3.1.2).

2.3.1.1 Unconscious Reasoning

In recent years, neuroscience has contributed a lot to the study of emotions through the development of novel methods for studying emotional processes and their neural correlates. In particular, new methods used in affective neuroscience, e.g., functional magnetic resonance imaging (FMRI), lesion studies, genetics, electro-physiology, paved the way towards the understanding of the neural circuitry that underlies emotional experience and of the manner in which emotional states influence health and life outcomes. A key contribution in the last two decades has been to provide evidence against the notion that emotions are subcortical and limbic, whereas cognition is cortical.

This notion was reinforcing the flawed Cartesian dichotomy between thoughts and feelings [97]. There is now ample evidence that the neural substrates of cognition and emotion overlap substantially [95]. Cognitive processes, such as memory encoding and retrieval, causal reasoning, deliberation, goal appraisal, and planning, operate continually throughout the experience of emotion. This evidence points to the importance of considering the affective components of any human-computer interaction [41]. Affective neuroscience, in particular, has provided evidence that elements of emotional learning can occur without awareness [229] and elements of emotional behavior do not require explicit processing [40]. Affective information processing mainly takes place at unconscious level (U-level) [119].

Reasoning, at this level, relies on experience and intuition, which allow for fast and effortless problem-solving. Hence, rather than reflecting upon various considerations in sequence, the U-level forms a global impression of the different issues. In addition, rather than applying logical rules or symbolic codes (e.g., words or numbers), the U-level considers vivid representations of objects or events. Such representations are laden with the emotions, details, features, and sensations that correspond to the objects or events.

Such human capability of summarizing huge amounts of inputs and outputs from previous situations, in order to find useful patterns that may work at the present time, is implemented here by means of AffectiveSpace. By reducing the dimensionality of the matrix representation of AffectNet, in fact, AffectiveSpace compresses the feature space of affective common-sense knowledge into one that allows for better gain a global insight and human-scale understanding. In cognitive science, the term 'compression' refers to transforming diffuse and distended conceptual structures that are less congenial to human understanding so they become better suited to our human-scale ways of thinking.

Compression is achieved hereby balancing the number of singular values discarded when synthesizing AffectiveSpace, in a way that the affective common-sense knowledge representation is neither too concrete nor too abstract with respect to the detail granularity needed for performing a particular task. The reasoning-by-analogy capabilities of AffectiveSpace, hence, are exploited at U-level to achieve digital intuition about the input data. In particular, the vector space representation of affective common-sense knowledge is clustered according the Hourglass model using the sentic medoids technique [58], in a way that concepts that are semantically and affectively related to the input data can be intuitively retrieved by analogy and unconsciously crop out to the C-level.

2.3.1.2 Conscious Reasoning

U-level and C-level are two conceptual systems that operate by different rules of inference. While the former operates emotionally and intuitively, the latter relies on logic and rationality. In particular, the C-level analyzes issues with effort, logic, and deliberation rather than relying on intuition. Hence, while at U-level the vector space representation of AffecNet is exploited to intuitively guess semantic and affective relations between concepts, at C-level associations between concepts are made according to the actual connections between different nodes in the graph representation of affective common-sense knowledge. Memory is not a 'thing' that is stored somewhere in a mental warehouse and can be pulled out and brought to the fore. Rather, it is a potential for reactivation of a set of concepts that together constitute a particular meaning. Associative memory involves the unconscious activation of networks of association–thoughts, feelings, wishes, fears, and perceptions that are connected, so that activation of one node in the network leads to activation of the others [325].

Sentic activation aims to implement such a process through the ensemble application of dimensionality-reduction and graph-mining techniques. Specifically, the semantically and affectively related concepts retrieved by means of AffectiveSpace at U-level are fed into AffectNet in order to crawl it according to how such seed concepts are interconnected to each other and to other concepts in the semantic network. To this end, spectral association [143] is employed. Spectral association is a technique that assigns values, or activations, to seed concepts and spreads their values across the AffectNet graph.

This operation, which is an approximation of many steps of spreading activation, transfers the most activation to concepts that are connected to the seed concepts by short paths or many different paths in affective common-sense knowledge. These related concepts are likely to have similar affective values. This can be seen as an alternate way of assigning affective values to all concepts, which simplifies the process by not relying on an outside resource such as WNA. In particular, a matrix A that relates concepts to other concepts, instead of their features, is built and the scores are added up over all relations that relate one concept to another, disregarding direction.

Applying A to a vector containing a single concept spreads that concept's value to its connected concepts. Applying A^2 spreads that value to concepts connected by two links (including back to the concept itself). But the desired operation is to spread the activation through any number of links, with diminishing returns, so the operator wanted is:

$$1 + A + \frac{A^2}{2!} + \frac{A^3}{3!} + \ldots = e^A \tag{2.11}$$

This odd operator, e^A, can be calculated because A can be factored. A is already symmetric, so instead of applying Lanczos' method [176] to AA^T and getting the SVD, it can be applied directly to A to obtain the spectral decomposition $A = V\Lambda V^T$. As before, this expression can be raised to any power and everything but the power of Λ cancelled. Therefore, $e^A = Ve^{\Lambda}V^T$. This simple twist on the SVD allows for the calculation of calculate spreading activation over the whole matrix instantly. As with the SVD, these matrices can be truncated to k axes and, therefore, space can be saved while generalizing from similar concepts. The matrix can also be rescaled so that activation values have a maximum of 1 and do not tend to collect in highly-connected concepts such as `person`, by normalizing the truncated rows of $Ve^{\Lambda/2}$ to unit vectors, and multiplying that matrix by its transpose to get a rescaled version of $Ve^{\Lambda}V^T$. Spectral association can spread not only positive, but also negative activation values. Hence, unconscious reasoning at U-level is exploited not only to retrieve concepts that are most semantically and affectively related, but also concepts that are most likely to be unrelated with the input data (lowest dot product).

While the former are exploited to spread semantics and sentics across the AffectNet graph, the latter are used to contain such an activation in a way that potentially unrelated concepts (and their twins) do not get triggered. This brain-inspired ensemble application of dimensionality-reduction and graph-mining techniques (herein after referred to as unconscious and conscious reasoning, respectively) allows sentic activation to more efficiently infer semantics and sentics from natural language text.

Sentic activation was tested on the benchmark for affective common-sense knowledge (BACK) by comparing concept classification results obtained by applying the AffectiveSpace process (U-level), spectral association (C-level), and the ensemble of U-level and C-level. Results showed that sentic activation achieves +13.9 % and +8.2 % accuracy than the AffectiveSpace process and spectral association, respectively.

2.3.2 Hourglass Model

The study of emotions is one of the most confused (and still open) chapters in the history of psychology. This is mainly due to the ambiguity of natural language, which does not facilitate the description of mixed emotions in an unequivocal way. Love and other emotional words like anger and fear, in fact, are suitcase words (many different meanings packed in), not clearly defined and meaning different things to different people [214].

Hence, more than 90 definitions of emotions have been offered over the past century and there are almost as many theories of emotion, not to mention a complex array of overlapping words in our languages to describe them. Some categorizations include cognitive versus non-cognitive emotions, instinctual (from the amygdala) versus cognitive (from the prefrontal cortex) emotions, and also categorizations based on duration, as some emotions occur over a period of seconds (e.g., surprise), whereas others can last years (e.g., love).

The James-Lange theory posits that emotional experience is largely due to the experience of bodily changes [157]. Its main contribution is the emphasis it places on the embodiment of emotions, especially the argument that changes in the bodily concomitants of emotions can alter their experienced intensity. Most contemporary neuroscientists endorse a modified James-Lange view, in which bodily feedback modulates the experience of emotion [94]. In this view, emotions are related to certain activities in brain areas that direct our attention, motivate our behavior, and determine the significance of what is going on around us. Pioneering works by Broca [35], Papez [241], and MacLean [200] suggested that emotion is related to a group of structures in the centre of the brain called limbic system (or paleomammalian brain), which includes the hypothalamus, cingulate cortex, hippocampi, and other structures. More recent research, however, has shown that some of these limbic structures are not as directly related to emotion as others are, while some non-limbic structures have been found to be of greater emotional relevance [182].

Philosophical studies on emotions date back to ancient Greeks and Romans. Following the early Stoics, for example, Cicero enumerated and organized the emotions into four basic categories: *metus* (fear), *aegritudo* (pain), *libido* (lust), and *laetitia* (pleasure). Studies on evolutionary theory of emotions, in turn, were initiated in the late nineteenth century by Darwin [98]. His thesis was that emotions evolved via natural selection and, therefore, have cross-culturally universal counterparts. In the early 1970s, Ekman found evidence that humans share six basic emotions: happiness, sadness, fear, anger, disgust, and surprise [115]. Few tentative efforts to detect non-basic affective states, such as fatigue, anxiety, satisfaction, confusion, or frustration, have been also made [70, 109, 164, 243, 258, 280] (Table 2.7).

Table 2.7 Some existing definition of basic emotions. The most widely adopted model for affect recognition is Ekman's, although is one of the poorest in terms of number of emotions (Source: [50])

Author	#Emotions	Basic emotions
Ekman	6	Anger, disgust, fear, joy, sadness, surprise
Parrot	6	Anger, fear, joy, love, sadness, surprise
Frijda	6	Desire, happiness, interest, surprise, wonder, sorrow
Plutchik	8	Acceptance, anger, anticipation, disgust, joy, fear, sadness, surprise
Tomkins	9	Desire, happiness, interest, surprise, wonder, sorrow
Matsumoto	22	Joy, anticipation, anger, disgust, sadness, surprise, fear, acceptance, shy, pride, appreciate, calmness, admire, contempt, love, happiness, exciting, regret, ease, discomfort, respect, like

In 1980, Averill put forward the idea that emotions cannot be explained strictly on the basis of physiological or cognitive terms. Instead, he claimed that emotions are primarily social constructs; hence, a social level of analysis is necessary to truly understand the nature of emotion [17]. The relationship between emotion and language (and the fact that the language of emotion is considered a vital part of the experience of emotion) has been used by social constructivists and anthropologists to question the universality of Ekman's studies, arguably because the language labels he used to code emotions are somewhat US-centric. In addition, other cultures might have labels that cannot be literally translated to English (e.g., some languages do not have a word for fear [276]). For their deep connection with language and for the limitedness of the emotional labels used, all such categorical approaches usually fail to describe the complex range of emotions that can occur in daily communication. The dimensional approach [232], in turn, represents emotions as coordinates in a multi-dimensional space.

For both theoretical and practical reasons, an increasing number of researchers like to define emotions according to two or more dimensions. An early example is Russell's circumplex model [275], which uses the dimensions of arousal and valence to plot 150 affective labels. Similarly, Whissell considers emotions as a continuous 2D space whose dimensions are evaluation and activation [326]. The evaluation dimension measures how a human feels, from positive to negative. The activation dimension measures whether humans are more or less likely to take some action under the emotional state, from active to passive. In her study, Whissell assigns a pair of values <activation, evaluation> to each of the approximately 9,000 words with affective connotations that make up her Dictionary of Affect in Language.

Another bi-dimensional model is Plutchik's wheel of emotions, which offers an integrative theory based on evolutionary principles [246]. Following Darwin's thought, the functionalist approach to emotions holds that emotions have evolved for a particular function, such as to keep the subject safe [129, 131]. Emotions are adaptive as they have a complexity born of a long evolutionary history and, although we conceive emotions as feeling states, Plutchik says the feeling state is part of a process involving both cognition and behavior and containing several feedback loops. In 1980, he created a wheel of emotions, which consisted of 8 basic emotions and 8 advanced emotions each composed of 2 basic ones. In such model, the vertical dimension represents intensity and the radial dimension represents degrees of similarity among the emotions.

Besides bi-dimensional approaches, a commonly used set for emotion dimension is the <arousal, valence, dominance> set, which is known in the literature also by different names, including <evaluation, activation, power> and <pleasure, arousal, dominance> [208]. Recent evidence suggests there should be a fourth dimension: Fontaine et al. reported consistent results from various cultures where a set of four dimensions is found in user studies, namely <valence, potency, arousal, unpredictability> [127]. Dimensional representations of affect are attractive mainly because they provide a way of describing emotional states that is more tractable than using words.

This is of particular importance when dealing with naturalistic data, where a wide range of emotional states occurs. Similarly, they are much more able to deal with non-discrete emotions and variations in emotional states over time [86], since in such cases changing from one universal emotion label to another would not make much sense in real life scenarios.

Dimensional approaches, however, have a few limitations. Although the dimensional space allows to compare affect words according to their reciprocal distance, it usually does not allow making operations between these, e.g., for studying compound emotions. Most dimensional representations, moreover, do not model the fact that two or more emotions may be experienced at the same time. Eventually, all such approaches work at word level, which makes them unable to grasp the affective valence of multiple-word concepts.

The Hourglass of Emotions [57] is an affective categorization model inspired by Plutchik's studies on human emotions [246]. It reinterprets Plutchik's model by organizing primary emotions around four independent but concomitant dimensions, whose different levels of activation make up the total emotional state of the mind. Such a reinterpretation is inspired by Minsky's theory of the mind, according to which brain activity consists of different independent resources and that emotional states result from turning some set of these resources on and turning another set of them off [214]. This way, the model can potentially synthesize the full range of emotional experiences in terms of Pleasantness, Attention, Sensitivity, and Aptitude, as the different combined values of the four affective dimensions can also model affective states we do not have a specific name for, due to the ambiguity of natural language and the elusive nature of emotions.

The main motivation for the design of the model is the concept-level inference of the cognitive and affective information associated with text. Such faceted information is needed, within sentic computing, for a feature-based sentiment analysis, where the affective common-sense knowledge associated with natural language opinions has to be objectively assessed. Therefore, the Hourglass model systematically excludes what are variously known as self-conscious or moral emotions, e.g., pride, guilt, shame, embarrassment, moral outrage, or humiliation

[181, 188, 281, 308]. Such emotions, in fact, present a blind spot for models rooted in basic emotions, because they are by definition contingent on subjective moral standards. The distinction between guilt and shame, for example, is based in the attribution of negativity to the self or to the act. So, guilt arises when you believe you have done a bad thing, and shame arises when thinking of yourself as a bad person.

This matters because, in turn, these emotions have been shown to have different consequences in terms of action tendencies. Likewise, an emotion such as *schadenfreude* is essentially a form of pleasure, but it is crucially different from pride or happiness because of the object of the emotion (the misfortune of another that is not caused by the self), and the resulting action tendency (do not express). However, since the Hourglass model currently focuses on the objective inference of affective information associated with natural language opinions, appraisal-based emotions are not taken into account within the present version of the model.

The Hourglass model (Fig. 2.16) is a biologically-inspired and psychologically-motivated model based on the idea that emotional states result from the selective activation/disactivation of different resources in the brain.

Each such selection changes how we think by changing our brain's activities: the state of anger, for example, appears to select a set of resources that help us react with more speed and strength while also suppressing some other resources that usually make us act prudently. Evidence of this theory is also given by several FMRI experiments showing that there is a distinct pattern of brain activity that occurs when people are experiencing different emotions. Zeki and Romaya, for example, investigated the neural correlates of hate with an FMRI procedure [339]. In their experiment, people had their brains scanned while viewing pictures of people they hated. The results showed increased activity in the medial frontal gyrus, right putamen, bilaterally in the premotor cortex, in the frontal pole, and bilaterally in the medial insula of the human brain. Also the activity of emotionally enhanced memory retention can be linked to human evolution [39]. During early development, in fact, responsive behavior to environmental events is likely to have progressed as a process of trial-and-error. Survival depended on behavioral patterns that were repeated or reinforced through life and death situations. Through evolution, this process of learning became genetically embedded in humans and all animal species in what is known as 'fight or flight' instinct [33].

The primary quantity we can measure about an emotion we feel is its strength. But, when we feel a strong emotion, it is because we feel a very specific emotion. And, conversely, we cannot feel a specific emotion like fear or amazement without that emotion being reasonably strong. For such reasons, the transition between different emotional states is modelled, within the same affective dimension, using the function $G(x) = 1 - \frac{1}{\sigma\sqrt{2\pi}}e^{-x^2/2\sigma^2}$ with $\sigma = 0.5$, for its symmetric inverted bell curve shape that quickly rises up towards the unit value (Fig. 2.17).

In particular, the function models how the level of activation of each affective dimension varies from the state of 'emotional void' (null value) to the state of 'heightened emotionality' (unit value). Justification for assuming that the Gaussian function (rather than a step or simple linear function) is appropriate for modeling the variation of emotion intensity is based on research into the neural and behavioral

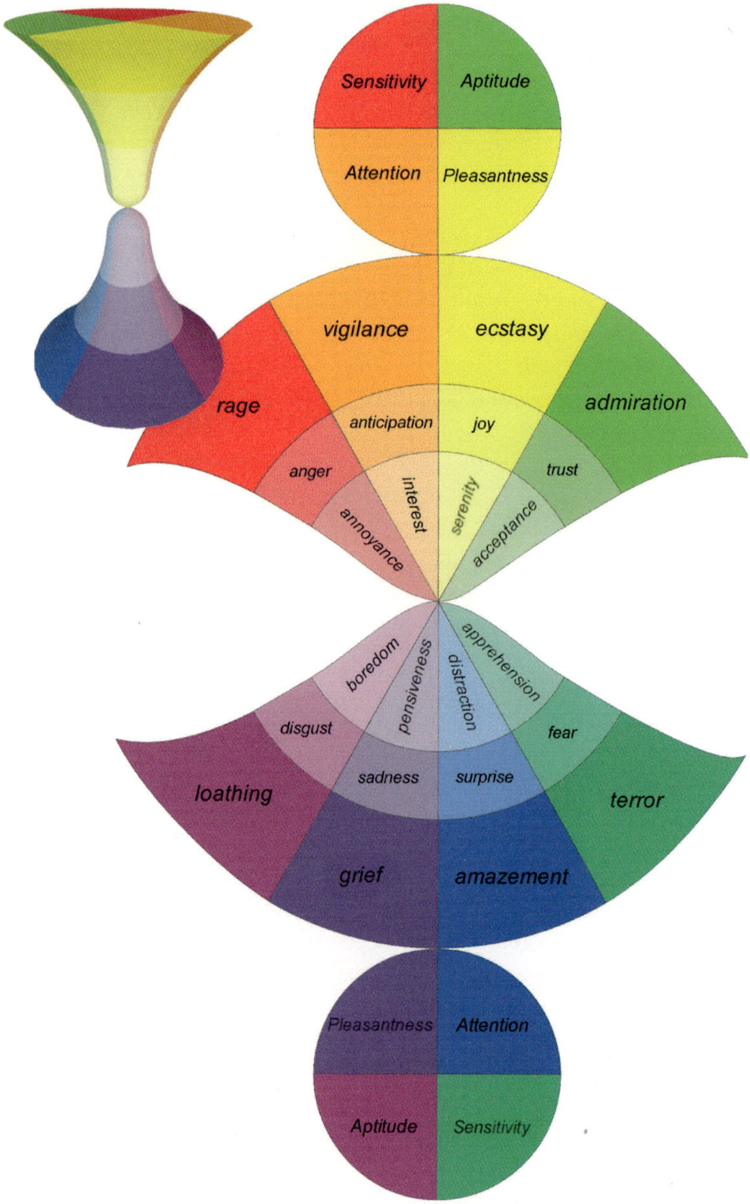

Fig. 2.16 The 3D model and the net of the Hourglass of Emotions. Since affective states go from strongly positive to null to strongly negative, the model assumes a hourglass shape (Source: [57])

correlates of emotion, which are assumed to indicate emotional intensity in some sense. In fact, nobody genuinely knows what function subjective emotion intensity follows, because it has never been truly or directly measured [22]. For example, the

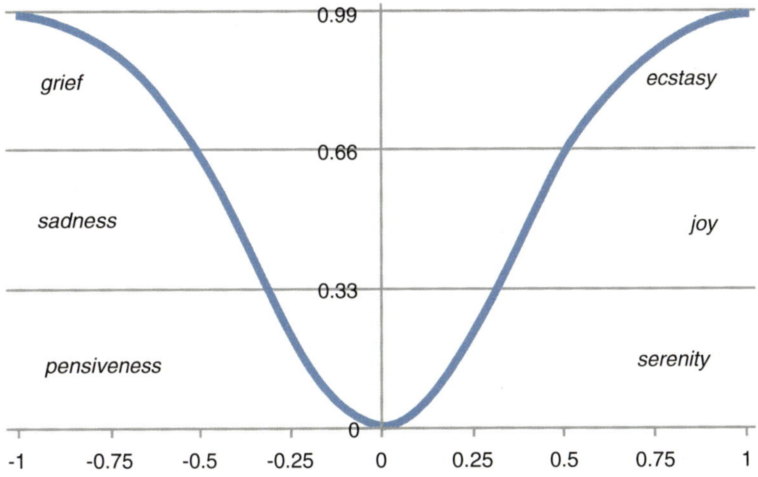

Fig. 2.17 The Pleasantness emotional flow. The passage from a sentic level to another is regulated by a Gaussian function that models how stronger emotions induce higher emotional sensitivity (Source: [57])

so-called Duchenne smile (a genuine smile indicating pleasure) is characterized by smooth onset, increasing to an apex, and a smooth, relatively lengthy offset [172].

More generally, Klaus Scherer has argued that emotion is a process characterized by non-linear relations among its component elements – especially physiological measures, which typically look Gaussian [189]. Emotions, in fact, are not linear [246]: the stronger the emotion, the easier it is to be aware of it. Mapping this space of possible emotions leads to a hourglass shape. It is worth noting that, in the model, the state of 'emotional void' is a-dimensional, which contributes to determine the hourglass shape. Total absence of emotion, in fact, can be associated with the total absence of reasoning (or, at least, consciousness) [92], which is not an envisaged mental state as, in the human mind, there is never nothing going on.

The Hourglass of Emotions, in particular, can be exploited in the context of HCI to measure how much respectively: the user is amused by interaction modalities (Pleasantness), the user is interested in interaction contents (Attention), the user is comfortable with interaction dynamics (Sensitivity), the user is confident in interaction benefits (Aptitude). Each affective dimension, in particular, is characterized by six levels of activation (measuring the strength of an emotion), termed 'sentic levels', which represent the intensity thresholds of the expressed or perceived emotion. These levels are also labeled as a set of 24 basic emotions [246], six for each of the affective dimensions, in a way that allows the model to specify the affective information associated with text both in a dimensional and in a discrete form (Table 2.8).

The dimensional form, in particular, is termed 'sentic vector' and is a four-dimensional *float* vector that can potentially synthesize the full range of emotional experiences in terms of Pleasantness, Attention, Sensitivity, and Aptitude. In the

Table 2.8 The sentic levels of the Hourglass model. Labels are organized into four affective dimensions with six different levels each, whose combined activity constitutes the 'total state' of the mind (Source: [50])

Interval	Pleasantness	Attention	Sensitivity	Aptitude
[G(1), G(2/3))	Ecstasy	Vigilance	Rage	Admiration
[G(2/3), G(1/3))	Joy	Anticipation	Anger	Trust
[G(1/3), G(0))	Serenity	Interest	Annoyance	Acceptance
(G(0), G(–1/3)]	Pensiveness	Distraction	Apprehension	Boredom
(G(–1/3), G(–2/3)]	Sadness	Surprise	Fear	Disgust
(G(–2/3), G(–1)]	Grief	Amazement	Terror	Loathing

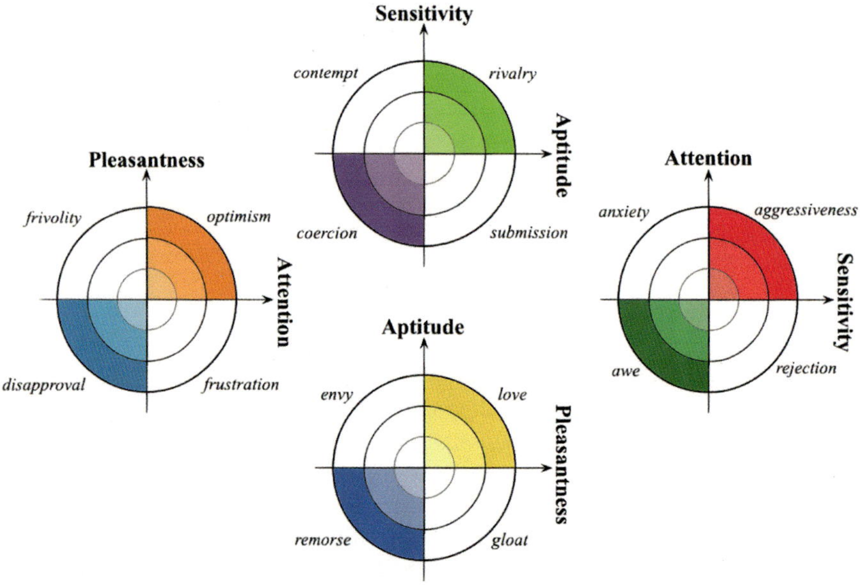

Fig. 2.18 Hourglass compound emotions of second level. By combining basic emotions pairwise, it is possible to obtain complex emotions resulting from the activation of two affective dimensions (Source: [57])

model, the vertical dimension represents the intensity of the different affective dimensions, i.e., their level of activation, while the radial dimension represents K-lines [212] that can activate configurations of the mind, which can either last just a few seconds or years. The model follows the pattern used in color theory and research in order to obtain judgements about combinations, i.e., the emotions that result when two or more fundamental emotions are combined, in the same way that red and blue make purple.

Hence, some particular sets of sentic vectors have special names, as they specify well-known compound emotions (Fig. 2.18). For example, the set of sentic vectors with a level of Pleasantness ∈ [G(2/3), G(1/3)), i.e., joy, a level of Aptitude ∈ [G(2/3), G(1/3)), i.e., trust, and a minor magnitude of Attention and Sensitivity,

Table 2.9 The second-level emotions generated by pairwise combination of the sentic levels of the Hourglass model. The co-activation of different levels gives birth to different compound emotions (Source: [50])

	Attention>0	Attention<0	Aptitude>0	Aptitude<0
Pleasantness>0	Optimism	Frivolity	Love	Gloat
Pleasantness<0	Frustration	Disapproval	Envy	Remorse
Sensitivity>0	Aggressiveness	Rejection	Rivalry	Contempt
Sensitivity<0	Anxiety	Awe	Submission	Coercion

are termed 'love sentic vectors' since they specify the compound emotion of love (Table 2.9). More complex emotions can be synthesized by using three, or even four, sentic levels, e.g., joy + trust + anger = jealousy.

Therefore, analogous to the way primary colors combine to generate different color gradations (and even colors we do not have a name for), the primary emotions of the Hourglass model can blend to form the full spectrum of human emotional experience. Beyond emotion detection, the Hourglass model is also used for polarity detection tasks. Since polarity is strongly connected to attitudes and feelings, in fact, it is defined in terms of the four affective dimensions, according to the formula:

$$p = \sum_{i=1}^{N} \frac{Pleasantness(c_i) + |Attention(c_i)| - |Sensitivity(c_i)| + Aptitude(c_i)}{3N} \tag{2.12}$$

where c_i is an input concept, N the total number of concepts, and 3 the normalization factor (as the Hourglass dimensions are defined as float $\in [-1, +1]$). In the formula, Attention is taken as absolute value since both its positive and negative intensity values correspond to positive polarity values (e.g., 'surprise' is negative in the sense of lack of Attention, but positive from a polarity point of view). Similarly, Sensitivity is taken as negative absolute value since both its positive and negative intensity values correspond to negative polarity values (e.g., 'anger' is positive in the sense of level of activation of Sensitivity, but negative in terms of polarity). The formula can be seen as one of the first attempts to show a clear connection between emotion recognition (sentiment analysis) and polarity detection (opinion mining).

2.3.3 *Sentic Neurons*

Affective analogical reasoning consists in processing the cognitive and affective information associated with natural language concepts, in order to compare the similarities between new and understood concepts and, hence, use such similarities to gain an understanding of the new concept. It is a form of inductive reasoning because it strives to provide understanding of what is likely to be true, rather than deductively proving something as fact. The reasoning process begins by determining the target concept to be learned or explained. It is then compared to a general matching concept whose semantics and sentics (that is, the conceptual and affective information associated with it) are already well-understood. The two concepts must be similar enough to make a valid, substantial comparison.

Affective analogical reasoning is based on the brain's ability to form semantic patterns by association. The brain may be able to understand new concepts more easily if they are perceived as being part of a semantic pattern. If a new concept is compared to something the brain already knows, it may be more likely that the brain will store the new information more readily.

Such a semantic association needs *high generalization performance*, in order to better match conceptual and affective patterns. Because of the dynamic nature of AffectiveSpace, moreover, affective analogical reasoning should be characterized by *fast learning speed*, in order for concept associations to be recalculated every time a new multi-word expression is inserted in AffectNet. Finally, the process should be of *low computational complexity*, in order to perform big social data analysis [66]. All such features are those typical of extreme learning machine (ELM), a machine learning technique that, in recent years, has proved to be a powerful tool to tackle challenging modeling problems [48, 151].

2.3.3.1 Extreme Learning Machine

The ELM approach [153] was introduced to overcome some well-known issues in back-propagation network [271] training, specifically, potentially slow convergence rates, the critical tuning of optimization parameters [320], and the presence of local minima that call for multi-start and re-training strategies. The ELM learning problem settings require a training set, X, of N labeled pairs, where $(\mathbf{x}_i, y_i)$, where $\mathbf{x}_i \in \mathscr{R}^m$ is the i-th input vector and $y_i \in \mathscr{R}$ is the associate expected 'target' value; using a scalar output implies that the network has one output unit, without loss of generality.

The input layer has m neurons and connects to the 'hidden' layer (having N_h neurons) through a set of weights $\{\hat{\mathbf{w}}_j \in \mathscr{R}^m; j = 1, \ldots, N_h\}$. The j-th hidden neuron embeds a bias term, $\hat{b}_j$,and a nonlinear 'activation' function, $\varphi(\cdot)$; thus the neuron's response to an input stimulus, $\mathbf{x}$, is:

$$a_j(\mathbf{x}) = \varphi(\hat{\mathbf{w}}_j \cdot \mathbf{x} + \hat{b}_j) \tag{2.13}$$

Note that (2.13) can be further generalized to a wider class of functions [152] but for the subsequent analysis this aspect is not relevant. A vector of weighted links, $\bar{\mathbf{w}}_j \in \mathscr{R}^{N_h}$, connects hidden neurons to the output neuron without any bias [150]. The overall output function, $f(\mathbf{x})$, of the network is:

$$f(\mathbf{x}) = \sum_{j=1}^{N_h} \bar{\mathbf{w}}_j a_j(\mathbf{x}) \tag{2.14}$$

It is convenient to define an 'activation matrix', $\mathbf{H}$, such that the entry $\{h_{ij} \in \mathbf{H}; i = 1, \ldots, N; j = 1, \ldots, N_h\}$ is the activation value of the j-th hidden neuron for the i-th input pattern. The $\mathbf{H}$ matrix is:

$$\mathbf{H} \equiv \begin{bmatrix} \varphi(\hat{\mathbf{w}}_1 \cdot \mathbf{x}_1 + \hat{b}_1) & \cdots & \varphi(\hat{\mathbf{w}}_{N_h} \cdot \mathbf{x}_1 + \hat{b}_{N_h}) \\ \vdots & \ddots & \vdots \\ \varphi(\hat{\mathbf{w}}_1 \cdot \mathbf{x}_N + \hat{b}_1) & \cdots & \varphi(\hat{\mathbf{w}}_{N_h} \cdot \mathbf{x}_N + \hat{b}_{N_h}) \end{bmatrix} \tag{2.15}$$

In the ELM model, the quantities $\{\hat{\mathbf{w}}_j, \hat{b}_j\}$ in (2.13) are set randomly and are not subject to any adjustment, and the quantities $\{\bar{\mathbf{w}}_j, \bar{b}\}$ in (2.14) are the only degrees of freedom. The training problem reduces to the minimization of the convex cost:

$$\min_{\{\bar{\mathbf{w}}, \bar{b}\}} \left\| \mathbf{H}\bar{\mathbf{w}} - \mathbf{y} \right\|^2 \tag{2.16}$$

A matrix pseudo-inversion yields the unique L_2 solution, as proven in [153]:

$$\bar{\mathbf{w}} = \mathbf{H}^{+}\mathbf{y} \tag{2.17}$$

The simple, efficient procedure to train an ELM therefore involves the following steps:

1. Randomly set the input weights $\hat{\mathbf{w}}_i$ and bias $\hat{b}_i$ for each hidden neuron;
2. Compute the activation matrix, $\mathbf{H}$, as per (2.15);
3. Compute the output weights by solving a pseudo-inverse problem as per (2.17).

Despite the apparent simplicity of the ELM approach, the crucial result is that even random weights in the hidden layer endow a network with a notable representation ability [153]. Moreover, the theory derived in [154] proves that regularization strategies can further improve its generalization performance. As a result, the cost function (2.16) is augmented by an L_2 regularization factor as follows:

$$\min_{\bar{\mathbf{w}}} \{ \left\| \mathbf{H}\bar{\mathbf{w}} - \mathbf{y} \right\|^2 + \lambda \left\| \bar{\mathbf{w}} \right\|^2 \} \tag{2.18}$$

2.3.3.2 The Emotion Categorization Framework

The proposed framework [45] is designed to receive as input, a natural language concept represented according to an M-dimensional space, and to predict the corresponding sentic levels for the four affective dimensions involved: Pleasantness, Attention, Sensitivity, and Aptitude. The dimensionality M of the input space stems from the specific design of AffectiveSpace. As for the outputs, in principle each affective dimension can be characterized by an analog value in the range $[-1, 1]$, which represents the intensity of the expressed or received emotion.

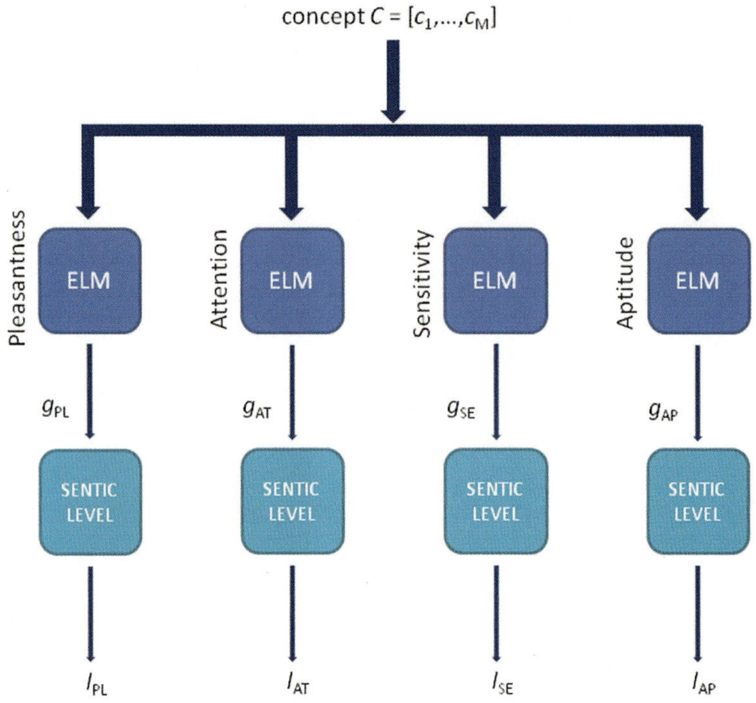

Fig. 2.19 The ELM-based framework for describing common-sense concepts in terms of the four Hourglass model's dimensions (Source: [253])

Indeed, those analog values are eventually remapped to obtain six different sentic levels for each affective dimension. The categorization framework spans each affective dimension separately, under the reasonable assumption that the various dimensions map perceptual phenomena that are mutually independent [50]. As a result, each affective dimension is handled by a dedicated ELM, which addresses a regression problem.

Thus, each ELM-based predictor is fed by the M-dimensional vector describing the concept and yields as output the analog value that would eventually lead to the corresponding sentic level. Figure 2.19 provides the overall scheme of the framework; here, g_X is the level of activation predicted by the ELM and l_X is the corresponding sentic level. In theory, one might also implement the framework showed in Fig. 2.19 by using four independent predictors based on a multi-class classification schema. In such a case, each predictor would directly yield as output a sentic level out of the six available. However, two important aspects should be taken into consideration. First, the design of a reliable multi-class predictor is not straightforward, especially when considering that several alternative schemata have been proposed in the literature without a clearly established solution. Second, the emotion categorization scheme based on sentic levels stem from an inherently analog model, i.e., the Hourglass of Emotions. This ultimately motivates the choice of designing the four prediction systems as regression problems.

In fact, the framework schematized in Fig. 2.19 represents an intermediate step in the development of the final emotion categorization system. One should take into account that every affective dimension can in practice take on seven different values: the six available sentic levels plus a 'neutral' value, which in theory correspond to the value $G(0)$ in the Hourglass model. In practice, though, the neutral level is assigned to those concepts that are characterized by a level activation that lies in an interval around $G(0)$ in that affective dimension. Therefore, the final framework should properly manage the eventual seven-level scale. To this end, the complete categorization system is set to include a module that is able to predict if an affective dimension is present or absent in the description of a concept. In the latter case, no sentic level should be associated with that affective dimension (i.e., $I_x = \text{null}$). This task is addressed here by exploiting the hierarchical approach presented in Fig. 2.20. Hence, given a concept and an affective dimension, first a SVM-based binary classifier is entitled to decide if a sentic level should be assessed. Accordingly, the ELM-based predictor is asked to assess the level of activation only if the SVM-based classifier determines that a sentic level should be associated with that concept. Otherwise, it is assumed that the neutral level should be associated with that concept (i.e., the corresponding affective dimension is not involved in the description of that concept). Obviously, such structure is replicated for each affective dimension. Figure 2.21 schematizes the complete framework.

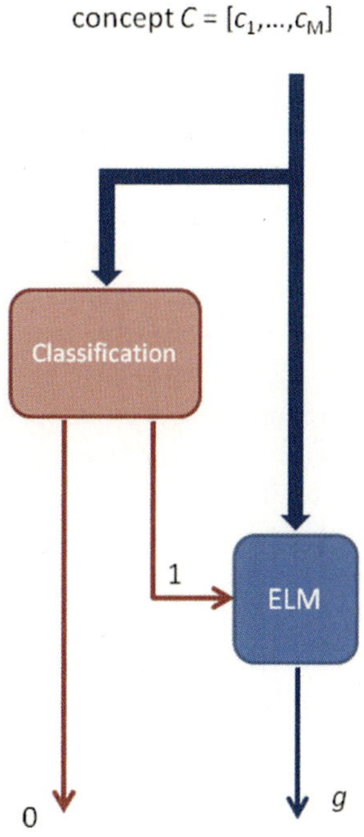

Fig. 2.20 The hierarchical scheme in which an SVM-based classifier first filters out unemotional concepts and an ELM-based predictor then classifies emotional concepts in terms of the involved affective dimension (Source: [253])

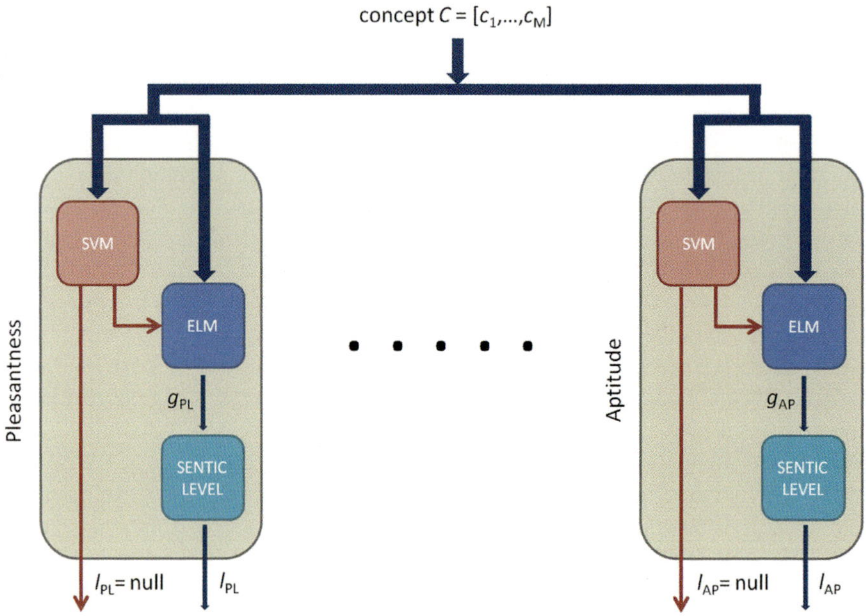

Fig. 2.21 The final framework: a hierarchical scheme is adopted to classify emotional concepts in terms of Pleasantness, Attention, Sensitivity, and Aptitude (Source: [253])

2.3.3.3 Experimental Results

The proposed emotion categorization framework has been tested both on a benchmark of 6,813 common-sense concepts and on a real-world dataset of 2,000 patient opinions. As for the benchmark, the Sentic API was used to obtain for each concept the corresponding sentic vector, i.e., the level of activation of each affective dimension. According to the Hourglass model, the Sentic API expresses the level of activation as an analog number in the range $[-1, 1]$, which are eventually mapped into sentic levels by adopting the Gaussian mapping function. Indeed, the neutral sentic level is codified by the value '0'. The format adopted by the Sentic API to represent the levels of activation actually prevents one to approach the prediction problem as an authentic regression task, as per Fig. 2.19.

The neutral sentic level corresponds to a single value in the analog range used to represent activations. Therefore, experimental results are presented as follows: firstly, the performance of the system depicted in Fig. 2.19 is analyzed (according to that set-up, the ELM-based predictors are not designed to assess the neutral sentic level); secondly, the performance of the complete framework (Fig. 2.21) is discussed; lastly, a use-case evaluation on the patient opinion dataset is proposed.

2.3.3.4 Accuracy in the Prediction of the Sentic Levels

The emotion categorization framework proposed in Fig. 2.19 exploits four independent ELM-based predictors to estimate the levels of activation of as many affective dimensions. In this experiment, it is assumed that each ELM-based predictor can always assess correctly a level of activation set to '0'. A cross-validation procedure has been used to robustly evaluate the performance of the framework.

As a result, the experimental session involved ten different experimental runs. In each run, 800 concepts randomly extracted from the complete benchmark provided the test set; the remaining concepts were heavenly split into a training set and a validation set. The validation set was designed to support the model selection phase, i.e., the selection of the best parameterization for the ELM predictors. In the present configuration, two quantities were involved in the model selection phase: the number of neurons N_h in the hidden layer and the regularization parameter λ.

The following parameters were used for model selection:

- $N_h \in [100, 1000]$ by steps of 100 neurons;
- $\lambda = \{1 \cdot 10^{-}6, 1 \cdot 10^{-}5, 1 \cdot 10^{-}4, 1 \cdot 10^{-}3, 1 \cdot 10^{-}2, 1 \cdot 10^{-}1, 1\}$.

In each run the performance of the emotion categorization framework was measured by using only the patterns included in the test set, i.e., the patterns that were not involved in the training phase or in the model selection phase. Table 2.10 reports the performance obtained by the emotion categorization framework over the ten runs. The table actually compares the results of three different sets up, which differs in the dimensionality M of AffectiveSpace that describe the concepts. Thus, Table 2.10 provides the results achieved with $M = 100$, $M = 70$, and $M = 50$.

The results refer to a configuration of the ELM predictors characterized by the following parameterization: $N_h = 200$ and $\lambda = 1$; such configuration was obtained by exploiting the model selection phase. The performance of each setting is evaluated according to the following quantities (expressed as average values over the ten runs):

- *Pearson's correlation coefficient*: the measure of the linear correlation between predicted levels of activation and expected levels of activation for the four predictors.
- *Strict accuracy*: the percentage of patterns for which the framework correctly predicted the four sentic levels; thus, a concept is assumed to be correctly

Table 2.10 Performance obtained by the emotion categorization framework over the ten runs with three different set-ups of AffectiveSpace (Source: [50])

	Correlation				Accuracy		
M	Pleasantness	Attention	Sensitivity	Aptitude	Strict	Smooth	Relaxed
100	0.69	0.67	0.78	0.72	39.4	73.4	87.0
70	0.71	0.67	0.78	0.72	41.0	75.4	88.4
50	0.66	0.66	0.77	0.71	40.9	75.3	86.4

classified only if the predicted sentic level corresponds to the expected sentic level for every affective dimension.

- *Smooth accuracy*: the percentage of patterns for which the framework correctly predicted three sentic levels out of four; thus, a concept is assumed to be correctly classified even when one among the four predictors fails to assign the correct sentic level.
- *Relaxed accuracy*: in this case, one relaxes the definition of correct prediction of the sentic level. As a result, given an affective dimension, the prediction is assumed correct even when the assessed sentic level and the expected sentic level are contiguous in Table 2.8. As an example, let suppose that the expected sentic level in the affective dimension Sensitivity for the incoming concept is 'annoyance'. Then, the prediction is assumed correct even when the assessed sentic level is 'anger' or 'apprehension'. Therefore, the relaxed accuracy gives the percentage of patterns for which the framework correctly predicted the four sentic levels according to such criterion.

In practice, the smooth accuracy and the relaxed accuracy allow one to take into account two crucial issues: the dataset can include noise and entries may incorporate a certain degree of subjectiveness. The results provided in Table 2.10 lead to the following comments:

- Emotion categorization is in fact a challenging problem; in this regard, the gap between strict accuracy and smooth/relaxed accuracies confirms that the presence of noise is a crucial issue.
- The ELM-based framework can attain satisfactory performance in terms of smooth accuracy and relaxed accuracy. Actually, the proposed framework scored a 75 % accuracy in correctly assessing at least three affective dimension for an input concept.
- Reliable performance can be achieved even when a 50-dimensional AffectiveSpace is used to characterize concepts. The latter result indeed represents a very interesting outcome, as previous approaches to the same problem in general exploited a 100-dimensional AffectiveSpace. In this respect, this analysis shows that the use of ELM-based predictors can reduce the overall complexity of the framework by shrinking the feature space.

2.3.3.5 Accuracy of the Complete Emotion Categorization System

The complete categorization system exploits the hierarchical approach presented in Fig. 2.20 to assess the level of activation of a concept. According to such a set-up, the accuracy of the SVM-based classifier is critical to the whole system's performance, as it handles the preliminary filtering task before that actual sentic description is evaluated. In principle, one might analyze the performance of the two components separately and assess the run-time generalization accuracy accordingly. Nevertheless, in the present context, the system performance has been measured as a whole, irrespectively of the internal structure of the evaluation scheme. On the other hand, one should also consider that, given a concept and a sentic dimension in which such concept should be assessed as neutral, to predict a low activation value is definitely less critical than predicting a large activation value.

Therefore, the system performance has been evaluated by avoiding considering as an error the cases in which the expected sentic level is 'neutral' and the assessed sentic level is the less intense (either positive or negative). As an example, given the sentic dimension Attention, to classify a neutral sentic level either as 'interest' or 'distraction' would not be considered an error. The performance of the framework has been evaluated by exploiting the same cross-validation approach already applied in the previous experimental session. In the present case, though, the model selection approach involved both the SVM-based classifiers and the ELM-based predictors. For the SVM classifiers, two quantities were set with model selection: the regularization parameter C and the width σ of the Gaussian kernel. The following parameters were used for model selection:

- $C = \{1, 10, 100, 1000\}$;
- $\sigma = \{0.1, 0.25, 0.5, 0.75, 1, 1.5, 2, 5, 10\}$.

The performance obtained by the framework over the ten runs was of 38.3 %, 72 %, and 79.8 %, for strict accuracy, smooth accuracy, and relaxed accuracy, respectively. In this case, the experimental session involved only the set-up $M = 50$, which already proved to attain a satisfactory trade-off between accuracy and complexity.

The results refer to a configuration of the SVM classifiers characterized by the following parameterization: $C = 1$ and $\sigma = 1.5$. As expected, the accuracy of the complete framework is slightly inferior to that of the system presented in the previous section. Indeed, the results confirm that the proposed approach can attain satisfactory accuracies by exploiting a 50-dimensional AffectiveSpace. In this regard, one should also notice that the estimated performance of the proposed methodology appears quite robust, as it is estimated on ten independent runs involving different compositions of the training and the test set.

Chapter 3
Sentic Patterns

Nature uses only the longest threads to weave her patterns, so that each small piece of her fabric reveals the organization of the entire tapestry.

Richard Feynman

Abstract This chapter introduces a novel framework for polarity detection that merges linguistics, common-sense computing, and machine learning. By allowing sentiments to flow from concept to concept based on the dependency relation of the input sentence, in particular, a better understanding of the contextual role of each concept within the sentence is achieved. This is done by means of a semantic parser, which extracts concepts from text, a set of linguistic patterns, which match specific structures in opinion-bearing sentences, and an extreme learning machine, which processes anything the patterns could not analyze for either lack of knowledge or constructions.

Keywords Semantic parsing • Linguistic patterns • Machine learning • Polarity detection • Ensemble classification

This chapter illustrates how SenticNet can be used for the sentiment analysis task of polarity detection (Fig. 3.1). In particular, a semantic parser is firstly used to deconstruct natural language text into concepts (Sect. 3.1). Secondly, linguistic patterns are used in concomitance with SenticNet to infer polarity from sentences (Sect. 3.2). If no match is found in SenticNet or in the linguistic patterns, machine learning is used (Sect. 3.3). Finally, the chapter proposes a comparative evaluation of the framework with respect to the state of the art in polarity detection from text (Sect. 3.4).

© Springer International Publishing Switzerland 2015

E. Cambria, A. Hussain, *Sentic Computing*, Socio-Affective Computing 1,
DOI 10.1007/978-3-319-23654-4_3

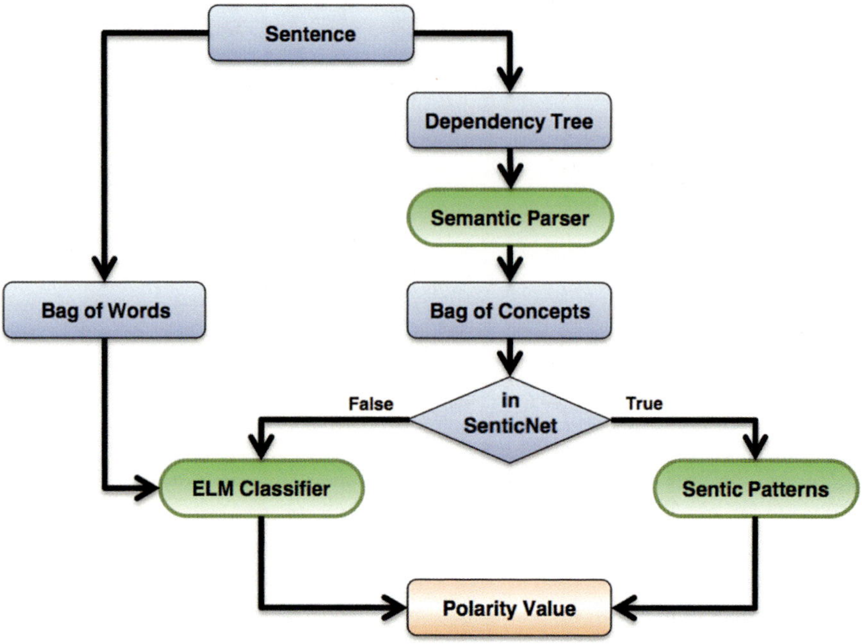

Fig. 3.1 Flowchart of the sentence-level polarity detection framework. Text is first decomposed into concepts. If these are found in SenticNet, sentic patterns are applied. If none of the concepts is available in SenticNet, the ELM classifier is employed (Source: The Authors)

3.1 Semantic Parsing

3.1.1 Pre-processing

Before text can be parsed, it needs to be normalized. To this end, a pre-processing module interprets all the affective valence indicators usually contained in opinionated text such as special punctuation, complete upper-case words, onomatopoeic repetitions, exclamation words, degree adverbs and emoticons. At the moment, this is done mainly by replacing fixed social expressions with their normalized version stored in a database of common patterns.

3.1.2 Concept Extraction

Concept extraction is about breaking text into clauses and, hence, deconstruct such clauses into bags of concepts, in order to feed these into a common-sense reasoning algorithm. For applications in fields such as real-time HCI and big social data analysis, in fact, deep natural language understanding is not strictly required: a sense of the semantics associated with text and some extra information (affect) associated with such semantics are often enough to quickly perform tasks such as emotion recognition and polarity detection.

3.1.2.1 From Sentence to Verb and Noun Chunks

The first step in the proposed algorithm breaks text into clauses. Each verb and its associated noun phrase are considered in turn, and one or more concepts is extracted from these. As an example, the clause "I went for a walk in the park", would contain the concepts `go_walk` and `go_park`.

The Stanford Chunker [202] is used to chunk the input text. A sentence like "I am going to the market to buy vegetables and some fruits" would be broken into "I am going to the market" and "to buy vegetables and some fruits". A general assumption during clause separation is that, if a piece of text contains a preposition or subordinating conjunction, the words preceding these function words are interpreted not as events but as objects. The next step of the algorithm then separates clauses into verb and noun chunks, as suggested by the following parse tree:

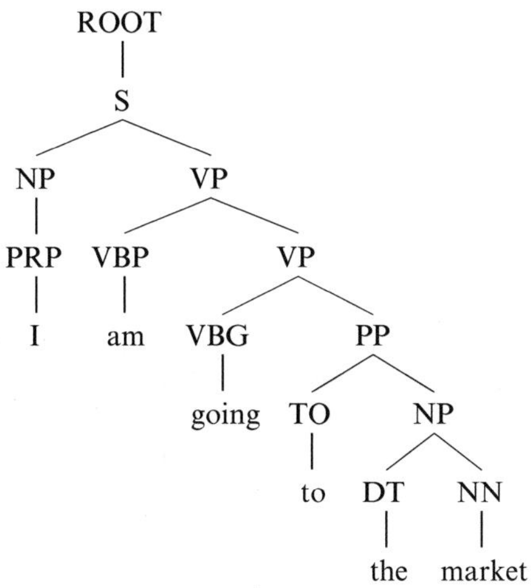

and

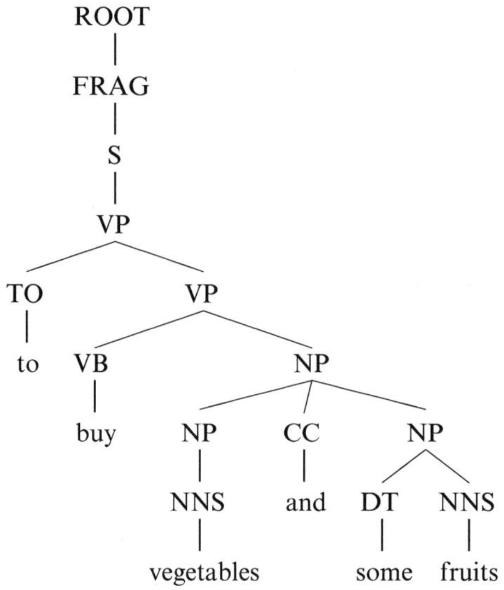

3.1.2.2 Obtaining the Full List of Concepts

Next, clauses are normalized in two stages. First, each *verb* chunk is normalized using the Stanford lemmatization algorithm. Second, each potential *noun* chunk associated with individual verb chunks is paired with the lemmatized verb in order to detect multi-word expressions of the form 'verb plus object'. Objects alone, however, can also represent a common-sense concept. To detect such expressions, a POS-based bigram algorithm checks noun phrases for stopwords and adjectives. In particular, noun phrases are first split into bigrams and then processed through POS patterns, as shown in Algorithm 1. POS pairs are taken into account as follows:

1. ADJECTIVE NOUN: The adj+noun combination and noun as a stand-alone concept are added to the objects list.
2. ADJECTIVE STOPWORD: The entire bigram is discarded.
3. NOUN ADJECTIVE: As trailing adjectives do not tend to carry sufficient information, the adjective is discarded and only the noun is added as a valid concept.
4. NOUN NOUN: When two nouns occur in sequence, they are considered to be part of a single concept. Examples include *butter scotch*, *ice cream*, *cream biscuit*, and so on.
5. NOUN STOPWORD: The stopword is discarded, and only the noun is considered valid.

Algorithm 1: POS-based bigram algorithm

```
Data: NounPhrase
Result: Valid object concepts
Split the NounPhrase into bigrams;
Initialize concepts to Null;
for each NounPhrase do
    while For every bigram in the NounPhrase do
        POS Tag the Bigram;
        if adj noun then
            add to Concepts: noun, adj+noun
        else if noun noun then
            add to Concepts: noun+noun
        else if stopword noun then
            add to Concepts: noun
        else if adj stopword then
            continue
        else if stopword adj then
            continue
        else
            Add to Concepts: entire bigram
        end
        repeat until no more bigrams left;
    end
end
```

6. STOPWORD ADJECTIVE: The entire bigram is discarded.
7. STOPWORD NOUN: In bigrams matching this pattern, the stopword is discarded and the noun alone qualifies as a valid concept.

The POS-based bigram algorithm extracts concepts such as `market`, `some_fruits`, `fruits`, and `vegetables`. In order to capture event concepts, matches between the object concepts and the normalized verb chunks are searched. This is done by exploiting a parse graph that maps all the multi-word expressions contained in the knowledge bases (Fig. 3.2).

Such an unweighted directed graph helps to quickly detect multi-word concepts, without performing an exhaustive search throughout all the possible word combinations that can form a common-sense concept.

Single-word concepts, e.g., `house`, that already appear in the clause as a multi-word concept, e.g., `beautiful_house`, in fact, are pleonastic (providing redundant information) and are discarded. In this way, the Algorithm 2 is able to extract event concepts such as `go_market`, `buy_some_fruits`, `buy_fruits`, and `buy_vegetables`, representing bags of concepts to be fed to a common-sense reasoning algorithm for further processing.

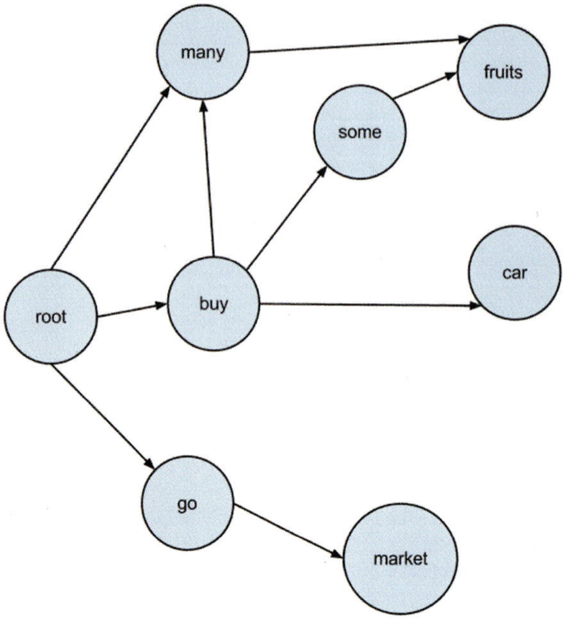

Fig. 3.2 Example parse graph for multi-word expressions (Source: [263])

3.1.3 Similarity Detection

Because natural language concepts may be expressed in a multitude of forms, it is necessary to have a technique for defining the similarity of multi-word expressions so that a concept can be detected in all its different forms.

The main aim of the proposed similarity detection technique, in fact, is to find concepts that are both syntactically and semantically related to the ones generated by the event concept extraction algorithm, in order to make up for concepts for which no matches are found in the knowledge bases. In particular, the POS tagging based

---

Algorithm 2: Event concept extraction algorithm

```
Data: Natural language sentence
Result: List of concepts
Find the number of verbs in the sentence;
for every clause do
    extract VerbPhrases and NounPhrases;
    lemmatize VERB;
    for every NounPhrase with the associated verb do
        find possible forms of objects;
        link all objects to lemmatized verb to get events;
    end
    repeat until no more clauses are left;
end
```

Algorithm 3: Finding similar concepts

```
Data: NounPhrase1, NounPhrase2
Result: True if the concepts are similar, else False
if Both phrases have atleast one noun in common then
    Objects1 := All Valid Objects for NounPhrase1;
    Objects2 := All Valid Objects for NounPhrase2;
    M1 = matches from KB for
    M1 := Ø ;
    M2 := Ø ;
    for all concepts in NounPhrase1 do
        M1 := M1 ∪ all property matches for concept;
    end
    for all concepts in NounPhrase2 do
        M2 := M2 ∪ all property matches for concept;
    end
    SetCommon = M1 ∪ M2;
    if length of SetCommon > 0 then
        The Noun Phrases are similar
    else
        They are not similar
    end
```

bigram algorithm is employed to calculate syntactic matches, while the knowledge bases are exploited to find semantic matches.

Beyond this, concept similarity may be exploited to merge concepts in the database and thus reduce data sparsity. When common-sense data is collected from different data sources, in fact, the same concepts tend to appear in different forms and merging these can be key for enhancing the common-sense reasoning capabilities of the system.

3.1.3.1 Syntactic Match Step

The syntactic match step checks whether two concepts have at least one object in common. For each noun phrase, objects and their matches from the knowledge bases are extracted, providing a collection of related properties for specific concepts. All the matching properties for each noun phrase are collected separately. The sets are then compared in order to identify common elements. If common elements exist, phrases are considered to be similar. Such similarity is deduced as shown in Algorithm 3.

3.1.3.2 Semantic Similarity Detection

Semantic similarity is calculated by means of AffectiveSpace and sentic medoids. In particular, in order to measure such semantic relatedness, AffectiveSpace is clustered by using a k-medoid approach [242]. Unlike the k-means algorithm (which does not pose constraints on centroids), k-medoids do assume that centroids must coincide with k observed points. The k-medoids approach is similar to the partitioning around medoids (PAM) algorithm, which determines a medoid for each cluster selecting the most centrally located centroid within that cluster.

Unlike other PAM techniques, however, the k-medoids algorithm runs similarly to k-means and, hence, requires a significantly reduced computational time. Given that the distance between two points in the space is defined as $D(e_i, e_j) = \sqrt{\sum_{s=1}^{d'} \left(e_i^{(s)} - e_j^{(s)}\right)^2}$, the adopted algorithm can be summarized as follows:

1. Each centroid $\bar{e}_i \in \mathbb{R}^{d'}$ $(i = 1, 2, \ldots, k)$ is set as one of the k most representative instances of general categories such as time, location, object, animal, and plant;
2. Assign each instance e_j to a cluster $\bar{e}_i$ if $D(e_j, \bar{e}_i) \leq D(e_j, \bar{e}_{i'})$ where $i(i') = 1, 2, \ldots, k$;
3. Find a new centroid $\bar{e}_i$ for each cluster c so that $\sum_{j \in Cluster\ c} D(e_j, \bar{e}_i) \leq \sum_{j \in Cluster\ c} D(e_j, \bar{e}_{i'})$;
4. Repeat step 2 and 3 until no changes are observed.

3.2 Linguistic Rules

The BoC model can represent the semantics associated with a natural language sentence much better than BoW. For example, a concept such as `cloud computing` would be split into two separate words, disrupting the semantics of the input sentence (in which, for example, the word `cloud` could wrongly activate concepts related to `weather`). The BoC model, however, would not be able to correctly infer the polarity of a sentence such as "the phone is nice but slow", in which it would just extract the concepts `phone`, `nice`, and `slow` (which in turn would be unlikely to result in a negative polarity on account of `nice` and `slow` bearing antithetic polarity values that nullify each other).

To this end, sentic patterns [249, 253] are further developed and applied. Sentic patterns are linguistic patterns for concept-level sentiment analysis, which allow sentiments to flow from concept to concept based on the dependency relation of the input sentence and, hence, to generate a binary (positive or negative) polarity value reflecting the feeling of the speaker (Fig. 3.3). It should be noted that, in some cases, the emotion attributed to a speaker can differ from his/her opinion.

For example, (1) conveys a negative sentiment, even though the speaker conveys that he/she is satisfied. There is a gap between the informational and emotional contents of the utterance and the aim of sentic patterns is extracting the latter.

(1) I am barely satisfied.

Similarly, a speaker can convey an objectively negative fact by presenting it in a positive way, as in (2).

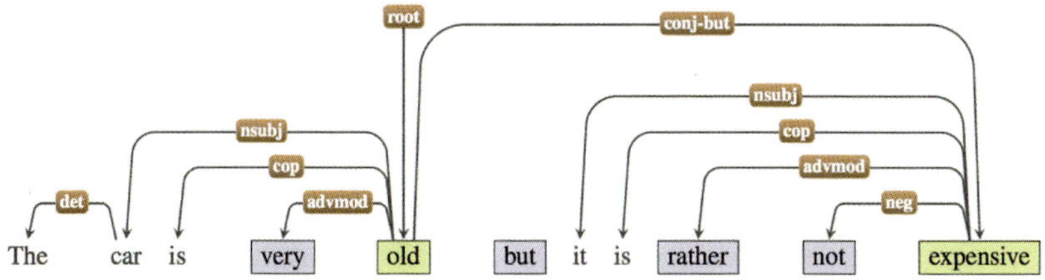

(a) Dependency tree of a sentence.

(b) The old way: averaging over a bag of sentiment words. The overall polarity of a sentence is given by the algebraic sum of the polarity values associated with each affect word divided by the total number of words.

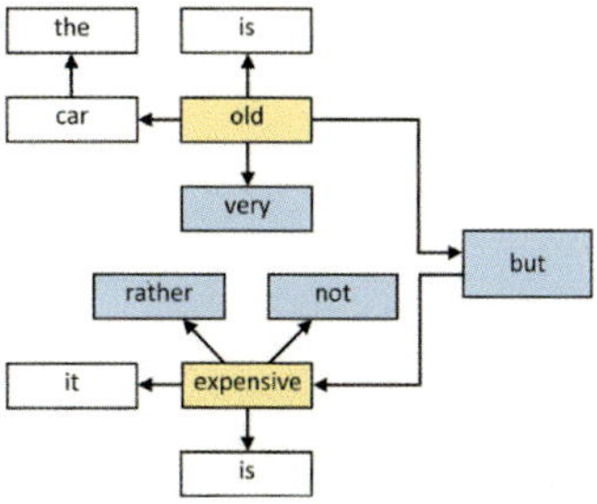

(c) The dependency tree of a sentence resembles an electronic circuit: words shown in blue can be thought as a sort of "boolean operations" acting on other words.

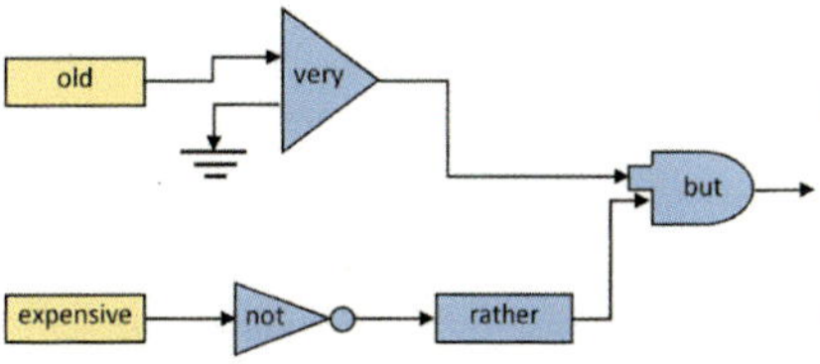

(d) The electronic circuit metaphor: sentiment words are "sources" while other words are "elements", e.g., *very* is an amplifier, *not* is a logical complement, *rather* is a resistor, *but* is an OR-like element that gives preference to one of its inputs.

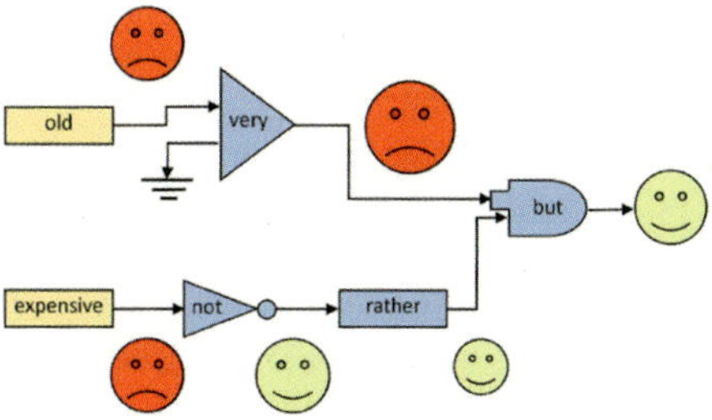

(e) The final sentiment data flow of the "signal" in the "circuit".

Fig. 3.3 The main idea behind sentic patterns: the structure of a sentence is like an electronic circuit where logical operators channel sentiment data-flows to output an overall polarity (Source: The Authors)

(2) It is fortunate that Paul died a horrible death.

Irrespective of Paul's fate, the (possibly psychotic) speaker presents it as a good thing. Hence, the inferred polarity is positive. Nevertheless, in most product or service reviews, the sentiment attributed to the speaker coincides with the opinion expressed. For example, if a sentence attributes a positive property to an object (e.g., "The battery is very good"), the sentiment of the speaker is considered corresponding to his/her evaluation.

In order to dynamically compute polarity, sentic patterns leverage on the ELM-based model for affective analogical reasoning [45] and on the syntactic dependency relations found in the input sentence. It is therefore an explicit approach that relies on linguistic considerations rather than on less interpretable models, such as those produced by most machine learning approaches. The upshot of this approach is that, besides being interpretable, it can take into account complex linguistic structures in a straightforward yet dynamic manner and can be easily modified and adapted.

The general template proposed for sentence-level polarity detection is illustrated in Sect. 3.2.1, notably by describing how polarity gets inverted (Sect. 3.2.1.2) and the way the calculus of polarity takes advantage of the discursive structure of the sentence (Sect. 3.2.1.3). The rules associated with specific dependency types are given in Sect. 3.2.2. A concrete example is given in Sect. 3.2.3.1.

3.2.1 General Rules

3.2.1.1 Global Scheme

The polarity score of a sentence is a function of the polarity scores associated with its sub-constituents. In order to calculate these polarities, sentic patterns consider each of the sentence's tokens by following their linear order and look at the dependency relations they have with other elements. A dependency relation is a binary relation characterized by the following features:

- The **type** of the relation that specifies the nature of the (syntactic) link between the two elements in the relation.
- The **head** of the relation: this is the element which is the pivot of the relation. Core syntactic and semantics properties (e.g., agreement) are inherited from the head.
- The **dependent** is the element that depends on the head and which usually inherits some of its characteristics (e.g., number, gender in case of agreement).

Most of the time, the active token is considered in a relation if it acts as the *head* of the relation, although some rules are an exception. Once the active token has been identified as the trigger for a rule, there are several ways to compute its contribution, depending on how the token is found in SenticNet. The preferred way is to consider the contribution not of the token alone, but in combination with the other element in the dependency relation.

This crucially exploits the fact that SenticNet is not just a polarity dictionary, but it also encodes the polarity of complex concepts. For example, in (3), the contribution of the verb `watch` will preferably be computed by considering the complex concept `watch_movie` rather than the isolated concepts `watch` and `movie`.

(3) I watched a movie.

If SenticNet has no entry for the multi-word concept formed by the active token and the element related to it, then the way individual contributions are taken into account depends on the type of the dependency relation. The specifics of each dependency type are given in Sect. 3.2.2.

Since SenticNet sometimes encodes sentiment scores for a token and a specific categorization frame, sentic patterns also check whether there is an entry for a frame corresponding to the active token and the part of speech of the other term in the dependency relation.

3.2.1.2 Polarity Inversion

Once the contribution of a token has been computed, sentic patterns check whether the token is in the scope of any polarity switching operator. The primary switching operator is negation: the use of negation on a positive token (4-a) yields a negative polarity (4-b).

(4) a. I liked the movie.
b. I did not like the movie.

However, double negation can keep the polarity of the sentence intact by flipping the polarity twice. For example, (5-a) is positive and (5-b) inverts its polarity. However, (5-c) keeps the polarity of (5-a) identical because in (5-c) *dislike* conveys negative polarity and, hence, nullifies the negation word *not.*

(5) a. I like it.
b. I do not like it.
c. I do not dislike it.

Besides negation, other polarity switching operators include:

- exclusives such as *only, just, merely*… [90]
- adverbs that type their argument as being low, such as *barely, hardly, least*…

(6) Paul is the least capable actor of his time.

- upper-bounding expressions like *at best, at most, less than*…
- specific constructions such as the use of past tense along with a comparative form of an adjective as in (7) or counter-factuals expressed by expressions like *would/could have been*

(7) a. My old phone was better. $\rightsquigarrow$ Negative
b. My old phone was slower. $\rightsquigarrow$ Positive

Whenever a token happens to be in the scope of such an element, its polarity score is inverted. Finally, inversion also happens when some specific scopeless expressions occur in a sentence, such as *except me*.

A shortcoming of this treatment of negation is that it does not take into account the different effects of negation on various layers of meaning. It is a well known fact in linguistics that some items convey complex meanings on different layers. Presupposition is probably the most studied phenomenon of this kind: both versions of (8) convey that John killed his wife, even though the second version is the negation of the first one [25, 165].

(8) a. John regrets killing his wife.
b. John does not regret killing his wife.

In the domain of sentiment related expressions, the class of *expressives* has comparable behavior, even though these elements have been analyzed as conventional implicatures rather than presuppositions [257]. For example, a verb like *waste* can be analyzed as conveying two distinct pieces of meaning: an event of money spending and a negative evaluation regarding this spending. In some cases, this negative component is not affected by negation: (9) convey that the phone is not worth the money, even though the verb *waste* is embedded under a negation.

(9) a. I will not waste my money on this phone.
b. I do not want to waste my money on this phone.
c. I did not waste my money on this phone.

Therefore, the current treatment of negation needs to be supplemented by classification of expressions indicating whether their negative (or positive) behavior has to be analyzed as a *main content*, affected by negation and other operators, or as a *projective content*, i.e., content that 'survives' or is non-canonically affected by operators that usually affect truth-conditional content. It might prove difficult to be exhaustive in this description since projection is not a purely semantic problem but is also affected by pragmatic contextual factors [286]. Nevertheless, it is conceivable to rely on a list of elements which convey sentiment on a clearly non-main level and to tune the algorithm to deal with them.

3.2.1.3 Coordinated and Discourse Structures

Coordination is an informationally rich structure for which sentic patterns have rules that do not specify which elements should be looked for in SenticNet, rather they indicate how the contributions of different elements should be articulated.

In some cases, a sentence is composed of more than one elementary discourse unit (in the sense of Asher and Lascarides [15]). In such cases, each unit is processed independently and the discourse structure is exploited in order to compute the overall polarity of the sentence, especially if an overt discourse cue is present.

At the moment, only structures that use an overt coordination cue are considered and the analysis is limited to adversative markers like *but* and to the conjunctions *and* and *or*.

But and Adversatives

Adversative items like *but*, *even though*, *however*, *although*, etc. have long been described as connecting two elements of opposite polarities. They are often considered as connecting two full-fledged discourse units in the majority of cases even when the conjuncts involve a form of ellipsis [269, 319].

Table 3.1 Adversative sentic patterns (Source: [253])

Left conjunct	Right conjunct	Total sentence
Pos.	Neg.	Neg.
Neg.	Pos.	Pos.
Pos.	Undefined	Neg.
Neg.	Undefined	Pos.
Undefined	Pos.	Pos.
Undefined	Neg.	Neg.

It has also long been observed that, in an adversative structure, the second argument "wins" over the first one [13, 332]. For example in (10-a) the overall attitude of the speaker goes against buying the car, whereas just inverting the order of the conjuncts yields the opposite effect (10-b) while keeping the informational content identical.

(10) a. This car is nice but expensive.
b. This car is expensive but nice.

Therefore, when faced with an adversative coordination, sentic patterns primarily consider the polarity of the right member of the construction for the calculation of the polarity of the overall sentence. If it happens that the right member of the coordination is unspecified for polarity, sentic patterns invert the polarity of the left member. The various possibilities are summarized in Table 3.1.

Specific heuristics triggered by tense are added to this global scheme. Whenever the two conjuncts share their topic and the second conjunct is temporally anterior to the first one, the overall polarity will be that of the first conjunct. Thus, in (11) since both conjuncts are about the director and the first one is posterior, the first one drives the polarity calculus.

(11) This director is making awful movies now, but he used to be good.

Another specific rule is implemented to deal with structures combining *not only* and *but also*, as in (12).

(12) The movie is not only boring but also offensive.

In such cases, *but* cannot be considered an opposition marker. Rather, both its conjuncts argue for the same goal. Therefore, when this structure is detected, the rule applied is the same as for conjunctions using *and* (cf. infra).

And

The conjunction *and* has been described as usually connecting arguments that have the same polarity and are partly independent [158]. Therefore, when a coordination with *and* is encountered, the overall polarity score of the coordination corresponds to the sum of both conjuncts. If only one happens to have a polarity score, this score is used with the addition of a small bonus to represent the fact that *and* connects independent arguments (i.e., the idea that speakers using *and* stack up arguments for their conclusions). In case of conflicts, the polarity of the second conjunct is used.

Or

A disjunction marked by *or* is treated in the same way as the *and* disjunction, i.e., by assuming that in the case where one of the conjuncts is underspecified, its polarity is determined by the other. However, there is no added bonus to the polarity score, since the semantics of disjunction do not imply independent arguments.

3.2.2 Dependency Rules

This section lists the whole set of rules that have been implemented to deal with specific dependency patterns. The main goal of these rules is to drive the way concepts are searched in SenticNet. One can roughly distinguish between two classes of dependencies:

- Relations of *complementation* where the dependent is an essential argument of the head.
- Relations of *modification* where the dependent is not sub-categorized by the head and acts as an adjunct.

Firstly, essential arguments of verbs (Sect. 3.2.2.1) will be treated, secondly modifiers (Sect. 3.2.2.2), and finally the rest of the rules (Sect. 3.2.2.3).

The default behavior of most rules is to build a multi-word concept formed by concatenating the concepts denoted by the head and the dependent of the relation (as exemplified in (3)). This multi-word concept is then searched in SenticNet. If it is not found, the behaviors of the rule differ.

Therefore, in the descriptions of the rules, it is systematically indicated:

- what triggers the rule;
- the behavior of the rule, i.e., the way it constructs complex concepts from the parts of the dependency relation under analysis.

To simplify the notation, the following notation is adopted:

- R denotes the relation type;
- h the head of the relation;
- d the dependent of the relation.

Therefore, writing $R(h, d)$ means that the head h has a dependency relation of type R with the dependent d. Typewriter font is used to refer to the concept denoted by a token, e.g., `movie` is the concept denoted by both tokens *movie* and *movies*. The concepts are the elements to be searched in SenticNet.

3.2.2.1 Relations of Complementation

Six relations of complementation, all centered on the verb as the head of the relation, are considered. One rule deals with the subject of the verb, the other three cover the different types of object a verb can take: noun phrases, adjective or full clauses.

Subject Nouns

Trigger: When the active token is found to be the syntactic subject of a verb.
Behavior: If the multi-word concept `(h,d)` is found in SenticNet, then it is used to calculate the polarity of the relation, otherwise the following strategies are followed:

- If the sentence is in passive voice and *h* and *d* are both negative, then the subject noun relation between *h* and *d* yields positive sentiment. If the sentence is not in passive voice, then the sentiment of the relation is negative.
- If *h* is negative and *d* is positive and the speaker is a first person, then the expressed sentiment is positive, otherwise sentic patterns predict a negative sentiment.
- If *h* is positive and *d* is negative, then the expressed sentiment is detected as negative by the sentic patterns.
- If *h* and *d* are both positive, then the relation results in a positive sentiment.

Example 1: In (13), *movie* is in a subject noun relation with *boring*.

(13) The movie is boring.

If the concept `(movie, boring)` is in SenticNet, its polarity is used. Otherwise, sentic patterns perform a detailed analysis of the relation to obtain the polarity. In this case, sentiment of `h` is treated as the sentiment of the relation.
Example 2: In (14), *relieve* is in a subject noun relation with *trouble*. Here, the polarity of *trouble* is negative and the polarity of *relieve* is positive. According to this rule, sentiment is carried by the *relieved*. So, here the sentence expresses a positive sentiment.

(14) His troubles were relieved.

Example 3: In (15), *success* is in subject noun relation with *pissed*. The polarity of *success* is positive while *pissed* has negative polarity. The final polarity of the sentence is negative according to this rule.

(15) My success has pissed him off.

Example 4: In (16), *gift* is in subject noun relation with *bad*. The polarity of *gift* is positive and *bad* is negative. Therefore, sentic patterns extract the polarity of the sentence as negative.

(16) Her gift was bad.

Direct Nominal Objects

This complex rule deals with direct nominal objects of a verb. Its complexity is due to the fact that the rule attempts to determine the modifiers of the noun in order to compute the polarity.

Trigger: When the active token is head verb of a direct object dependency relation.

Behavior: Rather than searching directly for the binary concept `(h,d)` formed by the head and the dependent, the rule first tries to find richer concepts by including modifiers of the nominal object. Specifically, the rule searches for relative clauses and prepositional phrases attached to the noun and if these are found, it searches for multi-word concepts built with these elements. Thus, if the dependent d is head of a relation of modification $R'(d, x)$, then sentic patterns will consider the ternary concept `(h,d,x)`. If this procedure fails and the binary concept `(h,d)` is not found either, the sign of the polarity is preferably driven by the head of the relation.

Example 1: In (17), sentic patterns first look for `(see,movie,in 3D)` in SenticNet and, if this is not found, they search for `(see,movie)` and then `(see, in 3D)`.

(17) Paul saw the movie in 3D.

`(movie,in 3D)` is not considered at this stage since it will be analyzed later under the standard rule for prepositional attachment. If the searching process fails, the polarity will be the one of `see` and eventually `movie`.

Example 2: In (18), first the concept `(make, pissed)` is searched in SenticNet and since it is not found, sentic patterns look for the polarity of *make* and *pissed* separately. As *make* does not exist in SenticNet, the polarity of *pissed* is considered as the polarity of the sentence (which is negative).

(18) You made me pissed off.

Example 3: In (19), the polarity of *love* is positive and the polarity of *movie* is negative as it is modified by a negative modifier *boring*. Sentic patterns set the polarity of this sentence as negative as the speaker says it is a boring movie though the subject *John* loves it.

(19) John loves this boring movie.

This rule has an exception when the subject is first person, i.e., the subject of the sentence and the speaker are the same.

Example 4: In (20), *hurt* has negative polarity and the polarity of *cat* is positive as it has a positive modifier *cute*. Thus, according to sentic patterns, the polarity of the sentence is negative.

(20) You have hurt the cute cat.

Complement Clause

This rule is fired when a sentence contains a finite clause which is subordinate to another clause: "That" and "Whether" are complement clauses.

Trigger: When a complement clause is found in a sentence.
Behavior: The sentence is split into two parts based on the complement clause:

- The sentiment expressed by the first part is considered as the final overall sentiment.
- If the first part does not convey any sentiment, then the sentiment of the second part is taken as the final sentiment.
- If the first part does not express any sentiment but a negation is present, then the sentiment of the second part is flipped.

Example 1: In (21), the sentiment expressed by the part of the sentence before "that" is positive, so the overall sentiment of the sentence is considered positive.

(21) I love that you did not win the match.

Example 2: In (22), the portion of the sentence before "whether" has no sentiment, but it contains a negation which alters the polarity of the second part. Thus, the overall polarity of the sentence becomes negative.

(22) I do not know whether he is good.

Adverbial Clause

Trigger: When a sentence contains an adverbial clause (i.e., "while").
Behavior: The role of "while" in a sentence is similar to the one of "but". Then, the sentic patterns first split the sentence into two parts by recognizing the subject and the use of comma in the sentence. Then, the overall sentiment of the sentence is conveyed by the second part.
Example: In (23), sentic patterns first identify the two parts of the sentence by recognizing the comma and the subject after the comma. The polarity of the first part (i.e., *i'm sure the quality of the product is fine*) is positive but the polarity of the second part (*the color is very different*) is neutral. Sentic patterns therefore detect the polarity of the sentence as negative.

(23) While I'm sure the quality of the product is fine, the color is very different.

Adjective and Clausal Complements

These rules deal with verbs having as complements either an adjective or a closed clause (i.e., a clause, usually finite, with its own subject).

Trigger: When the active token is head verb of one of the complement relations.
Behavior: First, sentic patterns look for the binary concept `(h,d)`. If it is found, the relation inherits its polarity properties. If it is not found:

- If both elements `h` and `d` are independently found in SenticNet, then the sentiment of `d` is chosen as the sentiment of the relation.
- If the dependent `d` alone is found in SenticNet, its polarity is attributed to the relation.

Example: In (24), *smells* is the head of a dependency relation with *bad* as the dependent.

(24) This meal smells bad.

The relation inherits the polarity of `bad`.

Open Clausal Complements

Open clausal complements are clausal complements of a verb that do not have their own subject, i.e., they usually share their subjects with the ones of the matrix clause.

Trigger: When the active token is the head predicate of the relation.[1]
Behavior: As for the case of direct objects, sentic patterns try to determine the structure of the dependent of the head verb. Here the dependent is itself a verb, therefore, sentic patterns attempt to establish whether a relation $R'(d,x)$ exists, where x is a direct object or a clausal complement of d. Sentic patterns are therefore dealing with three elements: the head/matrix verb (or predicate) h, the dependent predicate d, and the (optional) complement of the dependent predicate x. Once these have been identified, sentic patterns first test the existence of the ternary concept `(h,d,x)`. If this is found in SenticNet, the relation inherits its properties. If it is not found, sentic patterns check for the presence of individual elements in SenticNet.

- If `(d,x)` is found as well as `h` or if all three elements `h`, `d` and `x` are independently found in SenticNet, then the final sentiment score will be the one of `(d,x)` or it will be calculated from `d` and `x` by following the appropriate rule. The head verb affects the sign of this score. The rules for computing the sign are summarized in Table 3.2, where the final sign of the score is expressed as a function of the signs of the individual scores of each of the three relevant elements.
- If the dependent verb `d` is not found in SenticNet but the head verb `h` and the dependent's complement `x` can be found, then they are used to produce a score with a sign again corresponding to the rules stated in Table 3.2.

[1]Usually the token is a verb, although when the tensed verb is a copula, the head of the relation is rather the complement of the copula.

Table 3.2 Polarity algebra for open clausal complements (Source: [253])

Matrix predicate (h)	Dependent predicate (d)	Dep. comp. (x)	Overall polarity	Example
Pos	Pos	Pos	Pos	(25-a)
Pos	Pos	Neg	Neg	(25-b)
Pos	Neg	Pos	Neg	(25-c)
Pos	Neg	Neg	Pos	(25-d)
Neg	Pos	Pos	Neg	(25-e)
Neg	Pos	Neg	Neg	(25-f)
Neg	Neg	Pos	Neg	(25-g)
Neg	Neg	Neg	Neg	(25-h)
Pos	Neutral	Pos	Pos	(25-i)
Pos	Neutral	Neg	Neg	(25-j)
Neg	Neutral	Pos	Neg	(25-k)
Neg	Neutral	Neg	Neg	(25-l)

Example: In order to illustrate every case presented in Table 3.2, the paradigm in (25) is used. For each example, the final sign of the polarity is calculated according to Table 3.2. The examples assume the following:

- *h*, the matrix predicate, is either:
  - *perfect*, which has a positive polarity
  - *useless*, which has a negative polarity
- *d*, the dependent verb, is either:
  - *gain*, which has a positive polarity
  - *lose*, which has a negative polarity
  - *talk*, which is not found isolated in SenticNet, i.e., is considered neutral here
- *x*, the complement of the dependent verb, is either:
  - *money*, which has a positive polarity
  - *weight*, which has a negative polarity[2]

It must be remembered that for such examples it is assumed that the sentiment expressed by the speaker corresponds to his/her opinion on whatever *this* refers to in the sentence: if the speaker is positive about the thing he/she is talking about, it is considered that he/she is expressing positive sentiments overall.

[2]The negative score associated with *weight* does not reflect a deliberate opinion on the meaning of term. This score is extracted from SenticNet and has been automatically computed as explained in [61]. Thus, even though the term might not appear negative at first glance, its sentiment profile is nevertheless biased towards the negative.

(25) a. This is perfect to gain money.
b. This is perfect to gain weight.
c. This is perfect to lose money.
d. This is perfect to lose weight.
e. This is useless to gain money.
f. This is useless to gain weight.
g. This is useless to lose money.
h. This is useless to lose weight.
i. This is perfect to talk about money.
j. This is perfect to talk about weight.
k. This is useless to talk about money.
l. This is useless to talk about weight.

3.2.2.2 Modifiers

Modifiers, by definition, affect the interpretation of the head they modify. This explains why in most of the following rules the dependent is the guiding element for the computation of polarity.

Adjectival, Adverbial and Participial Modification

The rules for items modified by adjectives, adverbs or participles all share the same format.

Trigger: When the active token is modified by an adjective, an adverb or a participle.
Behavior: First, the multi-word concept `(h,d)` is searched in SenticNet. If it is not found, then the polarity is preferably driven by the modifier `d`, if it is found in SenticNet, otherwise `h`.
Example: In (26), both sentences involve elements of opposite polarities. The rule ensures that the polarity of the modifiers is the one that is used, instead of the one of the head of the relation: e.g., in (26-b) *beautifully* takes precedence over *depressed*.

(26) a. Paul is a bad loser.
b. Mary is beautifully depressed.

Unlike other NLP tasks such as emotion recognition, the main aim of sentiment analysis is to infer the polarity expressed by the speaker (i.e., the person who writes the review of a hotel, product, or service). Hence, a sentence such as (26-b) would be positive as it reflects the positive sentiment of the speaker.

Relative Clauses

Trigger: When the active token is modified by a relative clause, restrictive or not. The dependent is usually the verb of the relative clause.

Behavior: If the binary concept `(h,d)` is found in SenticNet, then it assigns polarity to the relation, otherwise the polarity is assigned (in order of preference):

- by the value of the dependent verb `d` if it is found in SenticNet.
- by the value of the active token `h` if it is found.

Example: In (27), *movie* is in relation with *love* which acts as a modifier in the relative clause.

(27) I saw the movie you love.

Assuming `(love, movie)` is not in SenticNet while `love` is, then the latter will contribute to the polarity score of the relation. If none of these is in SenticNet, then the dependency will receive the score associated with `movie`. In the case of (27), the polarity will be inherited at the top level because the main verb *see* is neutral. However, the overall polarity of a sentence like (28) is positive since, in case the subject is a first person pronoun, the sentence directly inherits the polarity of the main verb, here *like* (see Sect. 3.2.2.3 for more details).

(28) I liked the movie you love.

Similarly, (29) will obtain an overall negative sentiment because the main verb is negative.

(29) I disliked the movie you love.

Prepositional Phrases

Although prepositional phrases (PPs) do not always act as modifiers, we insert them in this section since the distinction is not significant for their treatment. Another reason is due to the fact that the Stanford dependency parser on which the framework relies does not differentiate between modifier and non-modifier PPs.

Trigger: The rule is activated when the active token is recognized as typing a prepositional dependency relation. In this case, the head of the relation is the element to which the PP attaches, and the dependent is the head of the phrase embedded in the PP. This means that the active element is not one of the two arguments of the relation but participates in the definition of its type.

Behavior: Instead of looking for the multi-word concept formed by the head *h* and the dependent *d* of the relation, sentic patterns use the preposition *prep* (corresponding to the active token) to build a ternary concept `(h, prep, d)`. If this is not found, then they search for the binary concept `(prep, d)` formed by the preposition and the dependent and use the score of the dependent `d` as a last tentative. This behavior is overridden if the PP is found to be a modifier of a noun phrase (NP) that acts as the direct object.

Example 1: In (30), the parser yields a dependency relation using *with* between the verb *hit* and the noun *hammer* (= the head of the phrase embedded in the PP).

(30) Bob hit Mary with a hammer.

Therefore, sentic patterns first look for the multi-word concept `(hit, with, hammer)` and, if this is not found, they look for `(with, hammer)` and finally `hammer` itself.

Example 2: In (31), the PP headed by *in* is a modifier of the verb *complete*, which is positive in SenticNet. *Terrible way* is however negative and, because it directly modifies the verb, the overall polarity is given by this element.

(31) Paul completed his work in a terrible way.

Example 3: In (32), the PP introduced by *in* is attached to the direct object of the predicate *is a failure*.

(32) This actor is the only failure in an otherwise brilliant cast.

Here, sentic patterns will ignore the contribution of the PP since the main sentiment is carried by the combination of the verb and its object, which is negative.

Adverbial Clause Modifier

This kind of dependency concerns full clauses that act as modifiers of a verb. Standard examples involve temporal clauses and conditional structures.

Trigger: The rule is activated when the active token is a verb modified by an adverbial clause. The dependent is the head of the modifying clause.

Behavior: If the binary concept `(h,d)` is found in SenticNet, then it is used for calculating the score. Otherwise, the rule assigns polarity by considering first the dependent `d`, then the head `h`.

Example: In (33), *playing* modifies *slows*. If the multi-word concept `(slow, play)` is not in SenticNet, then first `play` then `slow` will be considered.

(33) The machine slows down when the best games are playing.

Untyped Dependency

Sometimes the dependency parser detects two elements that keep a dependency relation but it is unable to type it properly. In this case, if the multi-word concept `(h,d)` is not found, the polarity is computed by considering the dependent `d` alone.

3.2.2.3 Other Rules

First Person Heuristics

On top of the rules presented so far, a specific heuristic for sentences having the first person pronoun as subject was implemented. In this case, the sentiment is essentially carried by the head verb of the relation. The contrast can be analyzed in (34):

(34) a. Paul likes bad movies.
b. I like bad movies.

Whereas (34-a) is a criticism of Paul and his tastes, (34-b) is speaker-oriented as he/she expresses his/her (maybe peculiar) tastes. What matters is that the speaker of (34-b) is being positive and uses the verb *like*. This overrides the calculus that would yield a negative orientation as in (34-a) by considering the combination of `like` and `bad movies`.

Similarly, in (35) the use of the first person overrides the effect produced by the relative clause *which you like*. The overall sentiment is entirely driven by the use of the verb *hate* which is negative.

(35) I hate the movie which you like.

Rule for the Preposition "Against"

In English, "against" is a preposition which carries a sentiment. Usually it is used as a negative sentiment expressing word. But, *against* can also be used in a sentence to express positive sentiment. Here, a few examples to explain the role of "against" in determining the sentiment of a sentence are given. In (36), *activity* has negative sentiment as it is modified by a negative modifier, i.e., *criminal*. Here, *against*, attached to the target *activity*, actually flips the polarity of *activity* and the overall sentiment of the sentence becomes positive.

(36) I am against all criminal activities.

In (37), *against* attaches to the target *love* which has positive polarity. Then, the overall sentiment of the sentence becomes negative.

(37) He is against me and your love.

If *against* attaches to a word with no polarity then the sentence sentiment turns negative.

3.2.3 Activation of Rules

The algorithm operates over the dependency parse tree of the sentence. Starting from the first (leftmost) relation in the tree, the rules corresponding to relations are activated: for a relation $R(A, B)$, the rules of the form R_i are activated to assign polarity (not necessarily the same) to the relation itself and to the words A and B. The rules for relations that involve either A or B are scheduled to be activated next; the main idea of the algorithm is taking into account the polarity already assigned to the relations and words previously processed. However, a rule may alter the order of activation of other rules if it needs additional information before it can proceed. For example, while computing the polarity of a relation $R(A, B)$, if A and B have any modifier, negation and subject-noun relation, then those relations are computed immediately. The reason is that such relations may alter the polarity of A and B. If there is no rule for a given relation $R(A, B)$, then it is left unprocessed and the new relations are scheduled for processing using the method described above.

When there are no relations scheduled for processing, the process restarts from the leftmost relation not yet processed for which a rule exists. The output of the algorithm is the polarity of the relation processed last. It accumulates the information of all relations in the sentence, because each rule takes into account the result of the previous ones, so that the information flows from the leftmost relation towards the rule executed last, which often corresponds to one of the rightmost relations. Below, for (38) the sentiment flow across the dependency arcs based on the sentic patterns is described.

(38) My failure makes him happy.

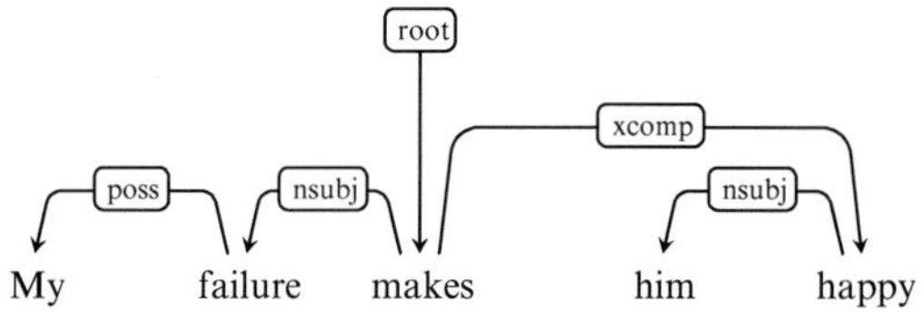

- First the relation between *my* and *failure* is considered. This is a *possession modifier* relation which does not satisfy any rule, so nothing has to be done.
- Then, the algorithm computes the polarity of the subject-noun relation between *make* and *failure*. The sentiment of this relation is negative according to the sentic patterns. The rule also assigns negative polarity to *make* which actually is a neutral word. This polarity is a contextual polarity to be used to compute the polarity of subsequent relations.
- Next, the polarity of the relation between *make* and *happy* is computed. This computation needs also the polarity of the relation computed in the previous step. Before computing the polarity of this relation, the subject-noun relation between *him* and *happy* is computed and a positive polarity is obtained. This polarity value does not alter the polarity of *happy*, which is positive according to SenticNet. *Make* has a negative polarity according to the previous step. Then,

there is a clausal complement relation between *make* and *happy*. Based on the clausal complement rule, sentic patterns assign negative polarity to this relation. After this computation there is no more relation left which satisfies the rules, so the sentence is assigned negative polarity by the algorithm.

(39) is another example to show the activation of rules and the flow of sentiments across the dependency arcs.

(39) You hurt the beautiful cat.

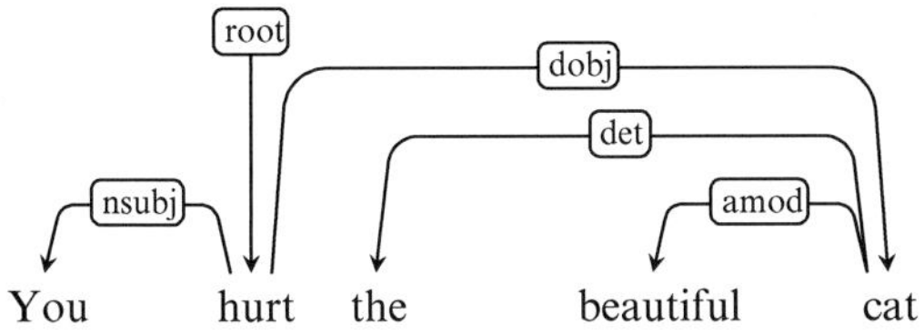

- First the algorithm encounters a subject-noun relation between *you* and *hurt*. As the polarity of *hurt* is negative, the algorithm assigns negative sentiment to the relation and *hurt* also maintains its negative polarity.
- Next, the algorithm finds *hurt* in a direct object relation with *cat*. To obtain the polarity of this relation, the algorithm first obtains the polarity of *cat* and the polarity of *hurt*, which was computed in the previous step. *Cat* does not exist in SenticNet but *cat* is modified by a positive word *beautiful*. So, *cat* is assigned positive polarity by sentic patterns. To compute the polarity of the direct object relation between *hurt* and *cat*, the algorithm has now all the necessary information. Based on the sentic patterns, it assigns negative polarity to this relation.
- The relation between *the* and *cat* does not satisfy any rule in sentic patterns. Nothing is done and there is no other relation to be processed. The final polarity of the sentence becomes negative.

3.2.3.1 Walking Through an Example

This section describes how the global sentiment for a complex example is computed. This is made in order to show how the sentiment flows in the treatment of a sentence. Figure 3.4 shows the parse tree for the sentence (40).

(40) The producer did not understand the plot of the movie inspired by the book and preferred to use bad actors.

The relevant dependency relations here are highlighted in Fig. 3.4. First, the discourse structure parser detects two discourse units conjoined by *and*. The final polarity will thus be a function of the elements π_1 = *The producer did not understand the plot of the movie based on the book* and π_2 = *[the producer] preferred to use bad actors*.

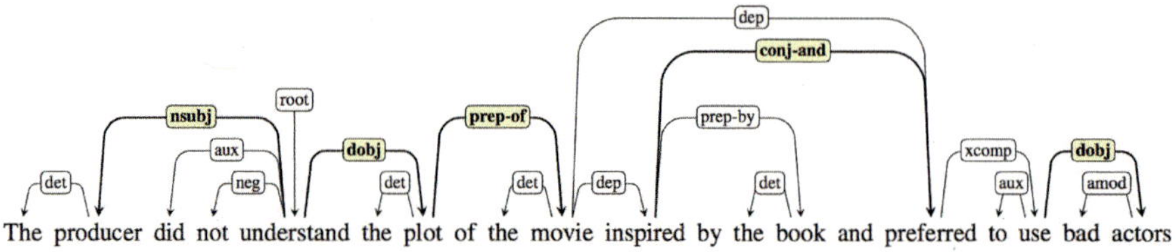

Fig. 3.4 Dependency tree for the sentence *The producer did not understand the plot of the movie inspired by the book and preferred to use bad actors* (Source: The Authors)

The computation of π_1 entails checking the relations in the following order:

- The subject relation *(understand, producer)* is considered to check whether the multi-word concept `(producer understand)` can be found in SenticNet. This is not the case, so nothing is done.
- The relations having the verb *understand* as their head are explored. Here there is only the direct object relation. In this relation the dependent object is modified in two ways: by a prepositional phrase and by a participial modifier

  Thus, sentic patterns will first try to find the multi-word concept `(understand, plot, of, movie)`. Since this one is not found, `(understand, plot, inspired)` is tried, and it is not in SenticNet either. Finally, sentic patterns fall back on the concept `(understand, plot)`, which is found in SenticNet. Therefore, the polarity stack is set at the corresponding positive value.
- Since the previous polarity is in the scope of a sentential negation, the sign of the previous score is switched to assign a negative value.

Now sentic patterns analyze π_2.

- The open clausal modification rule determines the dependent of the dependent. This case means identifying *actors* as the direct object of *use*.
- Since *actors* is modified by *bad*, it will inherit its negative orientation.
- The only relevant elements to computate the polarity, due to the open clausal complement, are *prefer* (positive) and *actor* (negative, because of its adjectival modification). Therefore, the final polarity score is also negative.

Finally, both the *and* conjuncts are negative, meaning that the overall polarity of the sentence is also negative with a value equal to the sum of the scores of each conjunct.

3.3 ELM Classifier

Despite being much more efficient than BoW and BoC models, sentic patterns are still limited by the richness of the knowledge base and the set of dependency-based rules. To be able to make a good guess even when no sentic pattern is matched or SenticNet entry found, the system resorts to machine learning. In particular, three well-known sentiment analysis datasets (Sect. 3.3.1), a set of four features per sentence (Sect. 3.3.2), and an artificial neural network (ANN) classifier (Sect. 3.3.3) are used to label text segments as positive or negative.

3.3.1 Datasets Used

3.3.1.1 Movie Review Dataset

The first dataset is derived from the benchmark corpus developed by Pang and Lee [236]. This corpus includes 1,000 positive and 1,000 negative movie reviews authored by expert movie reviewers, collected from rottentomatos.com, with all text converted to lowercase and lemmatized, and HTML tags removed. Originally, Pang and Lee manually labeled each review as positive or negative. Later, Socher et al. [293] annotated this dataset at sentence level. They extracted 11,855 sentences from the reviews and manually labeled them using a fine grained inventory of five sentiment labels: *strong positive*, *positive*, *neutral*, *negative*, and *strong negative*.

Since, this experiment is only about binary classification, sentences marked as neutral we removed and reduced the labels on the remaining sentences to positive or negative. Thus, the final movie dataset contained 9,613 sentences, of which 4,800 were labeled as positive and 4,813 as negative.

3.3.1.2 Blitzer Dataset

The second dataset is derived from the resource put together by Blitzer et al. [30], which consists of product reviews in seven different domains. For each domain there are 1,000 positive and 1,000 negative reviews. Only the reviews under the *electronics* category were used. From these 7,210 non-neutral sentences, 3505 sentences from positive reviews and 3,505 from negative ones were randomly extracted, and manually annotated as positive or negative. Note that the polarity of individual sentences does not always coincide with the overall polarity of the review: for example, some negative reviews contain sentences such as "This is a good product - sounds great", "Gets good battery life", "Everything you'd hope for in an iPod dock" or "It is very cheap".

3.3.1.3 Amazon Product Review Dataset

The reviews of 453 mobile phones from http://amazon.com were crawled. Each review was split into sentences, and each sentence then manually labeled by its sentiment labels. Finally, 115,758 sentences were obtained, out of which 48,680 were negative, 2,957 sentences neutral and 64,121 positive. In this experiment, only positive and negative sentences employed. So, the final *Amazon dataset* contained 112,801 sentences annotated as either positive or negative.

3.3.2 Feature Set

3.3.2.1 Common-Sense Knowledge Features

Common-sense knowledge features consist of concepts represented by means of AffectiveSpace. In particular, concepts extracted from text through the semantic parser are encoded as 100-dimensional real-valued vectors and then aggregated into a single vector representing the sentence by coordinate-wise summation: $x_i = \sum_{j=1}^{N} x_{ij}$, where x_i is the i-th coordinate of the sentence's feature vector, $i = 1, \ldots, 100$; x_{ij} is the i-th coordinate of its j-th concept's vector, and N is the number of concepts in the sentence.

3.3.2.2 Sentic Feature

The polarity scores of each concept extracted from the sentence were obtained from SenticNet and summed up to produce a single scalar feature.

3.3.2.3 Part-of-Speech Feature

The number of adjectives, adverbs, and nouns in the sentence; three separate features.

3.3.2.4 Modification Feature

This is a single binary feature. For each sentence, its dependency tree was obtained from the dependency parser. This tree was analyzed to determine whether there is any word modified by a noun, adjective, or adverb. The modification feature is set to 1 in case of any modification relation in the sentence; 0 otherwise.

3.3.2.5 Negation Feature

Similarly, the negation feature is a single binary feature determined by the presence of any negation in the sentence. It is important because the negation can invert the polarity of the sentence.

3.3.3 Classification

Sixty percent of the sentences were selected from each of the three datasets as the training set for the classification. The sentences from each dataset were randomly drawn in such a way to balance the dataset with 50 % negative sentences and 50 %

Table 3.3 Dataset to train and test ELM classifiers (Source: [253])

Dataset	Number of training sentences	Number of test sentences
Movie review dataset	5678	3935
Blitzer-derived dataset	4326	2884
Amazon dataset	67,681	45,120
Final dataset	77,685	51,939

Table 3.4 Performance of the classifiers: SVM/ELM classifier (Source: [253])

Training dataset	On movie review dataset	On Blitzer dataset	On Amazon dataset
Movie review	–	64.12 %/72.12 %	65.14 %/69.21 %
Blitzer	61.25 %/68.09 %	–	62.25 %/66.73 %
Amazon	69.77 %/70.03 %	72.23 %/73.30 %	–

Table 3.5 Feature analysis (Source: [253])

Features used	Accuracy (%)	Features used	Accuracy (%)
All	71.32	All except part-of-speech feature	70.41
All except common-sense	40.11	All except modification feature	71.53
All except sentic feature	70.84	All except negation feature	68.97

positive sentences. Again, ELM was used, which was found to outperform a state-of-the-art SVM in terms of both accuracy and training time. An overall 71.32 % accuracy was obtained on the Final Dataset described in Table 3.3 using ELM and 68.35 % accuracy using SVM. The classifiers were also trained on each single dataset and tested over all the other datasets. Table 3.4 reports the comparative performance results obtained in this experiment.

It can be noted from Table 3.4 that the model trained on the Amazon dataset produced the best accuracy compared to the movie review and Blitzer-derived datasets. For each of these experiments, ELM outperformed SVM. The best performance by the ELM classifier was obtained on the movie review dataset, while the SVM classifier performed best on the Blitzer dataset. The training and test set collected from different datasets are shown in Table 3.3.

Hence, whenever a sentence cannot be processed by SenticNet and sentic patterns, the ELM classifier makes a good guess about sentence polarity, based on the available features.

Although the ELM classifier has performed best when all features were used together, common-sense-knowledge based features resulted in the most significant ones. From the Table 3.5, it can be noticed that negation is also a useful feature. The other features were not found to have a significant role in the performance of the classifier but were still found to be useful for producing optimal accuracy. As ELM provided the best accuracy, Table 3.5 presents the accuracy of the ELM classifier. It should be noted that since the main purpose of this work is to demonstrate the ensemble use of linguistic rules, a detailed investigative study on features and their relative impact on ELM classifiers is proposed for future work, to further enrich and optimize the performance of the ensemble framework.

3.4 Evaluation

The polarity detection framework (available as a demo[3]) was tested on three datasets: the movie review dataset described in Sect. 3.3.1.1, the Blitzer-derived dataset described in Sect. 3.3.1.2 and the Amazon dataset described in Sect. 3.3.1.3. As shown by results below, the best accuracy is achieved when applying an ensemble of knowledge-based analysis and machine-learning classification, as the latter can act as reserve for the former when no match is found in SenticNet (Fig. 3.1). Table 3.6 shows a comparison of the experimental results.

3.4.1 Experimental Results

3.4.1.1 Results on the Movie Review Dataset

The proposed approach was evaluated on the movie review dataset and obtained an accuracy of 88.12 %, outperforming the state-of-the-art accuracy reported by Socher et al. [293] (85.40 %). Table 3.6 shows the results with ensemble classification and without ensemble classification. The table also presents a comparison of the proposed system with well-known state of the art. The table shows that the system performed better than [253] on the same movie review dataset. This is due to a new set of patterns and the use of a new training set for the ELM classifier, which helped to obtain better accuracy.

3.4.1.2 Results on the Blitzer-Derived Dataset

On the Blitzer-derived dataset described in Sect. 3.3.1.2, an accuracy of 88.27 % was achieved at the sentence level. The performance of the other benchmark sentiment-analysis systems was tested on this dataset. As on movie review dataset, the new

Table 3.6 Precision obtained using different algorithms on different datasets (Source: [253])

Algorithm	Movie review	Blitzer-derived	Amazon
RNN (Socher et al. [292])	80.00 %	–	–
RNTN (Socher et al. [293])	85.40 %	61.93 %	68.21 %
Poria et al. [253]	86.21 %	87.00 %	79.33 %
Sentic patterns	87.15 %	86.46 %	80.62 %
ELM classifier	71.11 %	74.49 %	71.29 %
Ensemble classification	**88.12 %**	**88.27 %**	**82.75 %**

[3]http://sentic.net/demo

patterns and new ELM training sets increased the accuracy over [253]. Further, the method by Socher et al. [293] was found to perform very poorly on the Blitzer dataset.

3.4.1.3 Results on the Amazon Dataset

The same table shows the results of sentic patterns on the Amazon dataset described in Sect. 3.3.1.3. Again, the proposed method outperforms the state-of-the-art approaches.

3.4.2 Discussion

The proposed framework outperforms the state-of-the-art methods on both the movie review and the Amazon datasets and shows even better results on the Blitzer-derived dataset. This shows that the framework is robust and not biased towards a particular domain. Moreover, while standard statistical methods require extensive training, both in terms of resources (training corpora) and time (learning time), sentic patterns are mostly unsupervised, except for the ELM module, which is, though, very fast, due to the use of ELM. The addition and improvement of the patterns, as noted in [253], has helped the system improve its results. Results show performance improvement over [253]. On the other hand, [293] has failed to obtain consistently good accuracy over both Blitzer and amazon datasets but obtained good accuracy over the movie review dataset. This is because the classifier proposed in [293] was trained on the movie review dataset only.

The proposed approach has therefore obtained a better accuracy than the baseline system. The three datasets described in Sects. 3.3.1.1, 3.3.1.2 and 3.3.1.3 were combined to evaluate the sentic patterns. From Sect. 3.3.1, the number of positive and negative sentences in the dataset can be calculated: this shows 72,721 positive and 56,903 negative sentences. If the system predicts all sentences as positive, this would give a baseline accuracy of 56.10 %. Clearly, the proposed system performed well above than the baseline system. It is worth noting that the accuracy of the system crucially depends on the quality of the output of the dependency parser, which relies on grammatical correctness of the input sentences. All datasets, however, contain ungrammatical sentences which penalize results. On the other hand, the formation of a balanced dataset for ELM classifiers actually has a strong impact on developing a more accurate classifier than the one reported in Poria et al. [253].

3.4.2.1 Effect of Conjunctions

Sentiment is often very hard to identify when sentences have conjunctions. The performance of the proposed system was tested on two types of conjunctions: *and*

Table 3.7 Performance of the proposed system on sentences with conjunctions and comparison with state-of-the-art (Source: [253])

System	AND (%)	BUT (%)
Socher et al. [293]	84.26	39.79
Poria et al. [253]	87.91	84.17
Extended sentic patterns	**88.24**	**85.63**

and *but*. High accuracy was achieved for both conjunctions. However, the accuracy on sentences containing *but* was somewhat lower as some sentences of this type do not match sentic patterns. Just over 27 % of the sentences in the dataset have *but* as a conjunction, which implies that the rule for *but* has a very significant impact on the accuracy. Table 3.7 shows the accuracy of the proposed system on sentences with *but* and *and* compared with the state of the art. The accuracy is averaged over all datasets.

3.4.2.2 Effect of Discourse Markers

Lin et al.'s [194] discourse parser was used to analyze the discourse structure of sentences. Out of the 1211 sentences in the movie review and the Blitzer dataset that contain discourse markers (*though*, *although*, *despite*), sentiment was correctly identified in 85.67 % sentences. According to Poria et al. [253], the discourse parser sometimes failed to detect the discourse structure of sentences such as *So, although the movie bagged a lot, I give very low rating*. Such problems were overcome by removing the occurrence of any word before the discourse marker when the marker occurred at either second or third position in the sentence.

3.4.2.3 Effect of Negation

With the linguistic rules from Sect. 3.2.1.2, negation was detected and its impact on sentence polarity was studied. Overall, 93.84 % accuracy was achieved on polarity detection from sentences with negation. Socher et al. [293] state that negation does not always reverse the polarity. According to them, the sentence "I do not like the movie" does not bear any negative sentiment, being neutral. For "The movie is not terrible," their theory suggests that this sentence does not imply that the movie is good, but rather that it is less bad, hence this sentence bears negative sentiment.

In the proposed annotation, this theory was not followed. The expression "not bad" was consider as implying satisfaction; thus, such a sentence was annotated as positive. Conversely, "not good" implies dissatisfaction and thus bears negative sentiment. Following this, the sentence "The movie is not terrible" is considered to be positive.

3.4.2.4 Examples of Differences Between the Proposed System and State-of-the-Art Approaches

Table 3.8 shows examples of various linguistic patterns and the performance of the proposed system across different sentence structures. Examples in Table 3.9 show that the proposed system produces consistent results on sentences carrying the same meaning although they use different words. In this example, the negative sentiment bearing word in the sentence is changed: in the first variant it is **bad**, in the second variant it is **bored**, and in the third variant it is **upset**. In each case, the system detects the sentiment correctly. This analysis also illustrates inconsistency of state-of-the-art approaches, given that the system [293] achieves the highest accuracy compared with other existing state-of-the-art systems.

Table 3.8 Performance comparison of the proposed system and state-of-the art approaches on different sentence structures (Source: [253])

Sentence	Socher et al. [293]	Sentic patterns
Hate iphone with a passion	Positive	Negative
Drawing has never been such easy in computer	Negative	Positive
The room is so small to stay	Neutral	Negative
The tooth hit the pavement and broke	Positive	Negative
I am one of the least happy people in the world	Neutral	Negative
I love starbucks but they just lost a customer	Neutral	Negative
I doubt that he is good	Positive	Negative
Finally, for the beginner there are not enough conceptual clues on what is actually going on	Positive	Negative
I love to see that he got injured badly	Neutral	Positive
I love this movie though others say it's bad	Neutral	Positive
Nothing can be better than this	Negative	Positive
The phone is very big to hold	Neutral	Negative

Table 3.9 Performance of the system on sentences bearing same meaning with different words (Source: [253])

Sentence	Socher et al. [293]	Sentic patterns
I feel **bad** when Messi scores fantastic goals	Neutral	Negative
I feel **bored** when Messi scores fantastic goals	Negative	Negative
I feel **upset** when Messi scores fantastic goals	Positive	Negative
I gave her a **gift**	Neutral	Positive
I gave her **poison**	Neutral	Negative

Table 3.10 Results obtained using SentiWordNet (Source: [253])

Dataset	Using SenticNet (%)	Using SentiWordNet (%)
Movie review	88.12	87.63
Blitzer	88.27	88.09
Amazon	82.75	80.28

3.4.2.5 Results Obtained Using SentiWordNet

An extensive experiment using SentiWordNet instead of SenticNet was carried out on all three datasets. The results showed SenticNet performed slightly better than SentiWordNet. A possible future direction of this work is the invention of a novel approach to combine SenticNet and SentiWordNet in the sentiment analysis framework. The slight difference in the accuracy reported in Table 3.10 confirmed that both the lexicons share similar knowledge but since SenticNet contains concepts, this helps increase accuracy.

For example, in the sentence "The battery lasts little", the proposed algorithm extracts the concept "last little" which exists in SenticNet but not in SentiWordNet. As a result, when SenticNet is used the framework labels the sentence with a "negative" sentiment but when using SentiWordNet the sentence is labeled with a "neutral" sentiment.

Chapter 4
Sentic Applications

The greatest mathematicians, as Archimedes, Newton, and Gauss, always united theory and applications in equal measure.

Felix Klein

Abstract This chapter lists a set of systems and applications that make use of SenticNet or sentic patterns (or both) for different sentiment analysis tasks. In particular, the chapter showcases applications in fields such as Social Web, human-computer interaction, and healthcare.

Keywords Troll filtering • Social media marketing • Photo management • Multi-modality • Healthcare

This chapter covers applications that makes use, in toto or in part, of SenticNet. Although SenticNet is a relatively new resource, there are a good number of works exploiting it for different sentiment analysis tasks. Xia et al. [335], for example, used SenticNet for contextual concept polarity disambiguation. In their approach, SenticNet was used as a baseline and contextual polarity was detected by a Bayesian method.

Other works [251, 254, 255] focused on extending or enhancing SenticNet. Poria et al. [254], for example, developed a fuzzy based SVM semi-supervised classifier to assign emotion labels to the SenticNet concepts. Several lexical and syntactic features as well as SenticNet based features were used to train the semi-supervised model.

Qazi et al. [261] used SenticNet for improving business intelligence from suggestive reviews. They built a supervised system where sentiment specific features were grasped from SenticNet (Fig. 4.1).

SenticNet can also be used for extracting concepts and discover domains from sentences. This use of SenticNet was studied by Dragoni et al. [110] proposed a fuzzy based framework which merges WordNet, ConceptNet and SenticNet to extract key concepts from a sentence. iFeel [14] is a system which allows its users to create their own sentiment analysis framework by combing SenticNet, SentiWordNet and other sentiment analysis methods.

© Springer International Publishing Switzerland 2015

E. Cambria, A. Hussain, *Sentic Computing*, Socio-Affective Computing 1,
DOI 10.1007/978-3-319-23654-4_4

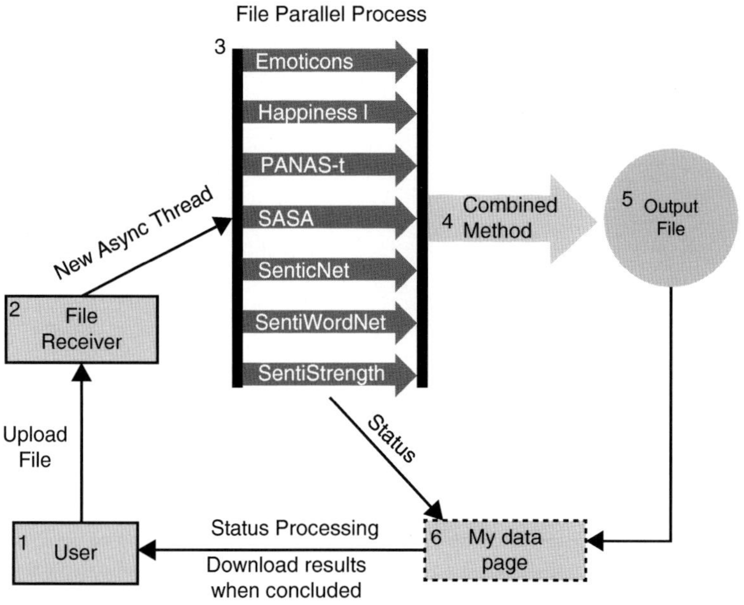

Fig. 4.1 iFeel framework (Source: [14])

SenticNet is also useful for crowd validation in e-health services as studied by [56]. Some approaches [333] focused on developing the multi-lingual concept level sentiment lexicon using the way SenticNet was built. SenticNet was also used to develop several supervised baseline methods [111, 132, 335]. Among other supervised approaches using SenticNet, the work by Chenlo et al. [77] is notable. They used SenticNet to extract bag of concepts and polarity features for subjectivity and sentiment analysis tasks. Chung et al. [85] used SenticNet concepts as seeds and proposed a method of random walk in the ConceptNet to retrieve more concepts along with polarity scores. Their method indeed aimed to expand SenticNet containing 265,353 concepts. After expanding SenticNet they formed Bag-of-Sentimental-Concepts features which is similar to Bag of Concepts features. Each dimension in the feature vector represents a concept and each concept is assigned a value by multiplying tf-idf and polarity value of the concept.

SenticNet has also been adopted for enhancing Twitter sentiment classification accuracy. The approach by Bravo et al. [34] used both SenticNet and SentiWordNet to improve the baseline Twitter classification system. On the other hand, Chikersal et al. [80] found out that SenticNet features are superior to SentiWordNet features for Twitter sentiment classification on Semeval dataset. SenticNet was also used for informal short text message (SMS) classification [132] and within a domain independent unsupervised sentiment-analysis system called Sentilo [265].

The rest of this section describes how sentic computing tools and techniques are employed for the development of applications in fields such as Social Web (Sect. 4.1), HCI (Sect. 4.2), and e-health (Sect. 4.3).

4.1 Development of Social Web Systems

With the rise of the Social Web, there are now millions of humans offering their knowledge online, which means information is stored, searchable, and easily shared. This trend has created and maintained an ecosystem of participation, where value is created by the aggregation of many individual user contributions. Such contributions, however, are meant for human consumption and, hence, hardly accessible and processable by computers. Making sense of the huge amount of social data available on the Web requires the adoption of novel approaches to natural language understanding that can give a structure to such data, in a way that they can be more easily aggregated and analyzed.

In this context, sentic computing can be exploited for NLP tasks requiring the inference of semantic and/or affective information associated with text, from big social data analysis [66] to management of online community data and metadata [135] to analysis of social network interaction dynamics [72]. This section, in particular, shows how the engine can be exploited for the development of a troll filtering system (Sect. 4.1.1), a social media marketing tool (Sect. 4.1.2), and an online personal photo management system (Sect. 4.1.3).

4.1.1 Troll Filtering

The democracy of the Web is what made it so popular in the past decades, but such a high degree of freedom of expression also gave birth to negative side effects – the so called 'dark side' of the Web. Be it real or virtual world, in fact, existence of malicious faction among inhabitants and users is inevitable. An example of this, in the Social Web context, is the exploitation of anonymity to post inflammatory, extraneous, or off-topic messages in an online community, with the primary intent of provoking other users into a desired emotional response or of otherwise disrupting normal on-topic discussion.

Such a practice is usually referred as 'trolling' and the generator of such messages is called 'a troll'. The term was first used in early 1990 and since then a lot of concern has been raised to contain or curb trolls. The trend of trolling appears to have spread a lot recently and it is alarming most of the biggest social networking sites since, in extreme cases such as abuse, has led some teenagers to commit suicide. These attacks usually address not only individuals, but also entire communities. For example, reports have claimed that a growing number of Facebook tribute pages had been targeted, including those in memory of the Cumbria shootings victims and soldiers who died in Afghanistan.

At present, users cannot do much other than manually delete abusive messages. Current anti-trolling methods, in fact, mainly consist in identifying additional accounts that use the same IP address and blocking fake accounts based on name and anomalous site activity, e.g., users who send lots of messages to non-friends or whose friend requests are rejected at a high rate. In July 2010, Facebook launched an application that gives users a direct link to advice, help, and the ability to report cyber problems to the child exploitation and online protection centre (CEOP). Reporting trouble through a link or a button, however, is too slow a process since social networking websites usually cannot react instantly to these alarms.

A button, moreover, does not stop users from being emotionally hurt by trolls and it is more likely to be pushed by people who actually do not need help rather than, for instance, children who are being sexually groomed and do not realize it. A prior analysis of the trustworthiness of statements published on the Web has been presented by Rowe and Butters [274]. Their approach adopts a contextual trust value determined for the person who asserted a statement as the trustworthiness of the statement itself. Their study, however, does not focus on the problem of trolling, but rather on defining a contextual accountability for the detection of web, email, and opinion spam.

The main aim of the troll filter [43] (Fig. 4.2) is to identify malicious contents in natural language text with a certain confidence level and, hence, automatically block trolls. To train the system, the concepts most commonly used by trolls are first

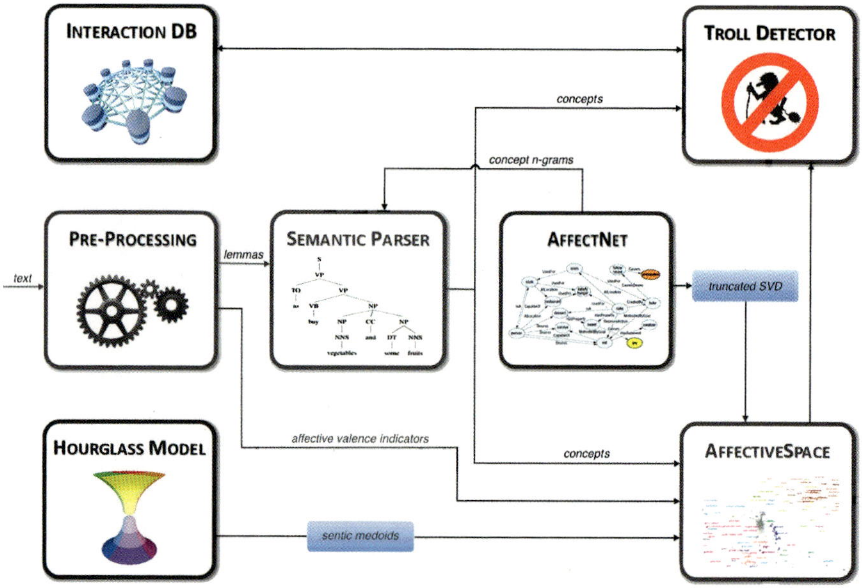

Fig. 4.2 Troll filtering process. Once extracted, semantics and sentics are used to calculate blogposts' level of trollness, which is then stored in the interaction database for the detection of malicious behaviors (Source: [50])

identified by using the CF-IOF technique and, then, this set is expanded through spectral association. In particular, after analyzing a set of 1000 offensive phrases extracted from Wordnik,[1] it was found that, statistically, a post is likely to be edited by a troll when its average sentic vector has a high absolute value of Sensitivity and a very low polarity. Hence, the *trollness* t_i associated with a concept c_i is defined as a float $\in [0, 1]$ such that:

$$t_i(c_i) = \frac{s_i(c_i) + |Sensitivity(c_i)| - p_i(c_i)}{3} \tag{4.1}$$

where s_i (float $\in [0, 1]$) is the semantic similarity of c_i with respect to any of the CF-IOF seed concepts, p_i (float $\in [-1, 1]$) is the polarity associated with the concept c_i, and 3 is the normalization factor. Hence, the total *trollness* of a post containing N concepts is defined as:

$$t = \sum_{i=1}^{N} \frac{3\ s_i(c_i) + 4\ |Sensitivity(c_i)| - Pleasantness(c_i) - |Attention(c_i)| - Aptitude(c_i)}{9N} \tag{4.2}$$

This information is stored, together with post type and content plus sender and receiver ID, in an interaction database that keeps trace of all the messages and comments interchanged between users within the same social network. Posts with a high level of *trollness* (current threshold has been set, using a trial-and-error approach, to 60 %) are labeled as troll posts and, whenever a specific user addresses more than two troll posts to the same person or community, his/her sender ID is labeled as troll for that particular receiver ID. All the past troll posts sent to that particular receiver ID by that specific sender ID are then automatically deleted from the website (but kept in the database with the possibility for the receiver to either visualize them in an apposite *troll folder* and, in case, restore them). Moreover, any new post with a high level of *trollness* edited by a user labeled as troll for that specific receiver is automatically blocked, i.e., saved in the interaction database but never displayed in the social networking website.

This information, encoded as a sentic vector, is given as input to a troll detector which exploits it, together with the semantic information coming directly from the semantic parser, to calculate the post's *trollness* and, eventually, to detect and block the troll (according to the information stored in the interaction database). As an example of troll filtering process output, a troll post recently addressed to the Indian author, Chetan Bhagat, can be considered: "You can't write, you illiterate douchebag, so quit trying, I say!!!". In this case, there are a very high level of Sensitivity (corresponding sentic level 'rage') and a negative polarity, which give a high percentage of *trollness*, as shown below:

[1] http://wordnik.com

<Concept: !'write'>
<Concept: 'illiterate'>
<Concept: 'douchebag'>
<Concept: 'quit try'>
<Concept: 'say'>
Semantics: 0.69
Sentics: [0.0, 0.17, 0.85, –0.43]
Polarity: –0.38
Trollness: 0.75

Because the approach adopted by Rowe and Butters [274] is not directly comparable with the developed troll filtering system, a first evaluation was performed by considering a set of 500 tweets manually annotated as troll and non-troll posts, most of which were fetched from Wordnik. In particular, true positives were identified as posts with both a positive troll-flag and a *trollness* $\in [0.6, 1]$, or posts with both a negative troll-flag and a *trollness* $\in [0, 0.6)$.

The threshold has been set to 60 % based on trial-and-error over a separate dataset of 50 tweets. Results show that, by using the troll filtering process, inflammatory and outrageous messages can be identified with good precision (82.5 %) and decorous recall rate (75.1 %). In particular, the F-measure value (78.9 %) is significantly high compared to the corresponding F-measure rates obtained by using IsaCore and AnalogySpace in place of the AffectiveSpace process (Table 4.1).

However, much better results are expected for the process evaluation at interaction level, rather than just at post level. In the future, in fact, the troll filtering process will be evaluated by monitoring not just single posts, but also users' holistic behavior, i.e., contents and recipients of their interaction, within the same social network.

4.1.2 Social Media Marketing

The advent of Web 2.0 made users more enthusiastic about interacting, sharing, and collaborating through social networks, online communities, blogs, wikis, and other online collaborative media. In the last years, this collective intelligence has spread to many different areas in the Web, with particular focus on fields related to our everyday life such as commerce, tourism, education, and health. The online review of commercial services and products, in particular, is an action that users usually

Table 4.1 Precision, recall, and F-measure values relative to the troll filter evaluation. The AffectiveSpace process performs consistently better than IsaCore and AnalogySpace in detecting troll posts (Source: [50])

Metric	IsaCore (%)	AnalogySpace (%)	AffectiveSpace (%)
Precision	57.1	69.1	82.5
Recall	40.0	56.6	75.1
F-measure	47.0	62.2	78.6

perform with pleasure, to share their opinions about services they have received or products they have just bought, and it constitutes immeasurable value for other potential buyers.

This trend opened new doors to enterprises that want to reinforce their brand and product presence in the market by investing in online advertising and positioning. In confirmation of the growing interest in social media marketing, several commercial tools have been recently developed to provide companies with a way to analyze the blogosphere on a large scale in order to extract information about the trend of the opinions relative to their products. Nevertheless most of the existing tools and the research efforts are limited to a polarity evaluation or a mood classification according to a very limited set of emotions. In addition, such methods mainly rely on parts of text in which emotional states are explicitly expressed and, hence, they are unable to capture opinions and sentiments that are expressed implicitly.

To this end, a novel social media marketing tool has been proposed [46] to provide marketers with an IUI for the management of social media information at semantic level, able to capture both opinion polarity and affective information associated with UGCs. A polarity value associated with an opinion, in fact, sometimes can be restrictive. Enriching automatic analysis of social media with affective labels such as 'joy' or 'disgust' can help marketers to have a clearer idea of what their customers think about their products. In particular, YouTube was selected as a social media source since, with its over two billions views per day, 24 h of video uploaded every minute, and 15 min a day spent by the average user, it represents more than 40 % of the online video market.[2] Specifically, the focus was on video reviews of mobile phones because of the quantity and the quality of the comments usually associated with them.

The social media analysis is performed through three main steps: firstly, comments are analyzed using the opinion-mining engine; secondly, the extracted information is encoded on the base of different web ontologies; finally, the resulting knowledge base is made available for browsing through a multi-faceted classification website. Social Web resources represent a peculiar kind of data that is characterized for a deeply interconnected nature. The Web itself is, in fact, based on links that bind together different data and information, and community-contributed multimedia resources characterize themselves for the collaborative way in which they are created and maintained.

An effective description of such resources therefore needs to capture and manage such interconnected nature, allowing to encode information not only about the resource itself, but also about the linked resources into an interconnected knowledge base. Encoding information relative to a market product to analyze its market trends represents a situation in which this approach is particularly suitable and useful. In this case, it is necessary not only to encode the information relative to product features, but also the information about the producer, the consumers, and their opinions.

[2]http://viralblog.com/research/youtube-statistics

The proposed framework for opinion description and management aims to be applicable to most online resources (videos, images, text) coming from different sources, e.g., online video sharing services, blogs, and social networks. To such purpose, it is necessary to standardize as much as possible the descriptors used in encoding the information about multimedia resources and people to which the opinions refer (considering that every website uses its own vocabulary), in order to make it univocally interpretable and suitable to feed other applications. For this reason, the information relative to multimedia resources and people is encoded using the descriptors provided by OMR[3] (Ontology for Media Resources) and FOAF[4] (Friend of a Friend Ontology), respectively. OMR represents an important effort to help circumventing the current proliferation of audio/video metadata formats, currently carried on by the W3C Media Annotations Working Group. It offers a core vocabulary to describe media resources on the Web, introducing descriptors such as 'title', 'creator', 'publisher', 'createDate', and 'rating', and it defines semantic-preserving mappings between elements from existing formats in order to foster the interoperability among them.

FOAF represents a recognized standard in describing people, providing information such as their names, birthdays, pictures, blogs, and especially other people they know, which makes it particularly suitable for representing data that appear in social networks and communities. OMR and FOAF together supply most of the vocabulary needed for describing media and people; other descriptors are added only when necessary. For example, OMR does not currently supply vocabulary for describing comments, which are analyzed here to extract the affective information relative to media. Hence, the ontology is extended by introducing the 'Comment' class and by defining for it the 'author', 'text', and 'publicationDate' properties.

In HEO, properties to link emotions to multimedia resources and people were introduced. In particular, 'hasManifestationInMedia' and 'isGeneratedByMedia' were defined to describe emotions that occur and are generated in media, respectively, while the property 'affectPerson' was defined to connect emotions to people. Additionally, WNA was exploited as an ontology in order to improve the hierarchical organization of emotions in HEO. Thus, the combination of HEO with WNA, OMR, and FOAF provides a complete framework to describe not only multimedia contents and the users that have created, uploaded, or interacted with them, but also the opinions and the affective content carried by the media and the way they are perceived by web users (Fig. 4.3).

As mentioned above, due to the way they are created and maintained, community-contributed multimedia resources are very different from standard web data. One fundamental aspect is the collaborative way in which such data is created, uploaded, and annotated. A deep interconnection emerges in the nature of these data and metadata, allowing for example to associate videos of completely different genre, but uploaded by the same user, or different users, even living in

[3]http://w3.org/TR/mediaont-10

[4]http://www.foaf-project.org

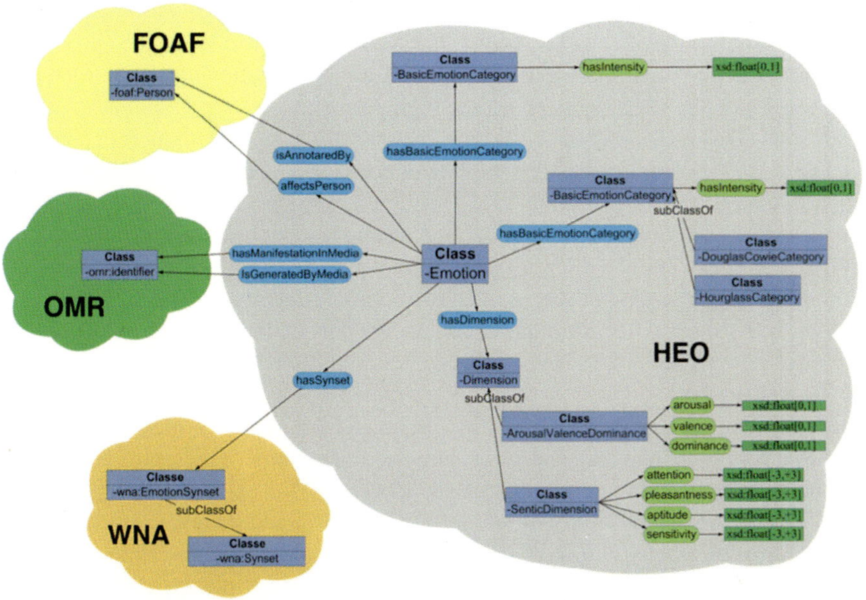

Fig. 4.3 Merging different ontologies. The combination of HEO, WNA, OMR and FOAF provides a comprehensive framework for the representation of social media affective information (Source: [50])

opposite sides of the world, who have appreciated the same pictures. In the context of social media marketing, this interdependence can be exploited to find similar patterns in customer reviews of commercial products and, hence, to gather useful information for marketing, sales, public relations, and customer service. Online reviews of electronic products, in particular, usually offer substantial and reliable information about the perceived quality of the products because of the size of the online electronics market and the type of customers related to it.

To visualize this information, the multi-faceted categorization paradigm is exploited. Faceted classification allows the assignment of multiple categories to an object, enabling the classifications to be ordered in multiple ways, rather than in a single, pre-determined, and taxonomic order. This makes possible to perform searches combining the textual approach with the navigational one.

Faceted search enables users to navigate a multi-dimensional information space by concurrently writing queries in a text box and progressively narrowing choices in each dimension. For this application, specifically, the SIMILE Exhibit API[5] is used. Exhibit consists of a set of Javascript files that allow for the creation of rich interactive web pages including maps, timelines, and galleries, with very detailed client-side filtering. Exhibit pages use the multi-faceted classification paradigm to display semantically structured data stored in a Semantic Web aware format, e.g.,

[5]http://simile-widgets.org/exhibit

RDF or JavaScript object notation (JSON). One of the most relevant aspects of Exhibit is that, once the page is loaded, the web browser also loads the entire data set in a lightweight database and performs all the computations (sorting, filtering, etc.) locally on the client-side, providing high performance.

Because they are one of the most prolific types of electronic products in terms of data reviews available on the Web, mobile phones were selected as a review target. In particular, a set of 220 models was considered. Such models were ranked as the most popular according to Kelkoo,[6] a shopping site featuring online shopping guides and user reviews, from which all the available information about each handset, such as model, brand, input type, screen resolution, camera type, standby time, and weight, was parsed. This information was encoded in RDF and stored in a Sesame[7] triple-store, a purpose-built database for the storage and retrieval of RDF metadata. YouTube Data API was then exploited to retrieve from YouTube database the most relevant video reviews for each mobile phone and their relative metadata such as duration, rating, upload date and name, gender, and country of the uploaders.

The comments associated with each video were also extracted and processed by means of sentic computing for emotion recognition and polarity detection. The extracted opinions in RDF/XML were then encoded using the descriptors defined by HEO, WNA, OMR, and FOAF, and inserted into the triple-store. Sesame can be embedded in applications and used to conduct a wide range of inferences on the information stored, based on RDFS and OWL type relations between data. In addition, it can also be used in a standalone server mode, much like a traditional database with multiple applications connecting to it. In this way, the knowledge stored inside Sesame can be easily queried; optionally, results can also be retrieved in a semantic aware format and used for other applications.

For the developed demo, the information contained in the triple-store was exported into a JSON file, in order to make it available for being browsed as a unique knowledge base through Exhibit interface. In the IUI, mobile phones are displayed through a dynamic gallery that can be ordered according to different parameters, e.g., model, price, and rating, showing technical information jointly with their video reviews and the opinions extracted from the relative comments (Fig. 4.4). By using faceted menus, moreover, it is possible to explore such information both using the search box (to perform keyword-based queries) and filtering the results using the faceted menus (by adding or removing constraints on the facet properties).

In this way, it becomes very easy and intuitive to search for mobile phones of interest: users can specify the technical features required using the faceted menus and compare different phones that match such requirements by consulting the video reviews and the opinions extracted from the relative comments. In addition, it is possible to explore in detail the comments of each video review through a specific Exhibit page in which comments are organized in a timeline and highlighted in different colors, according to the value of their polarity. Moreover, faceted menus allow filtering the comments according to the reviewers' information, e.g., age,

[6]http://kelkoo.co.uk

[7]http://openrdf.org

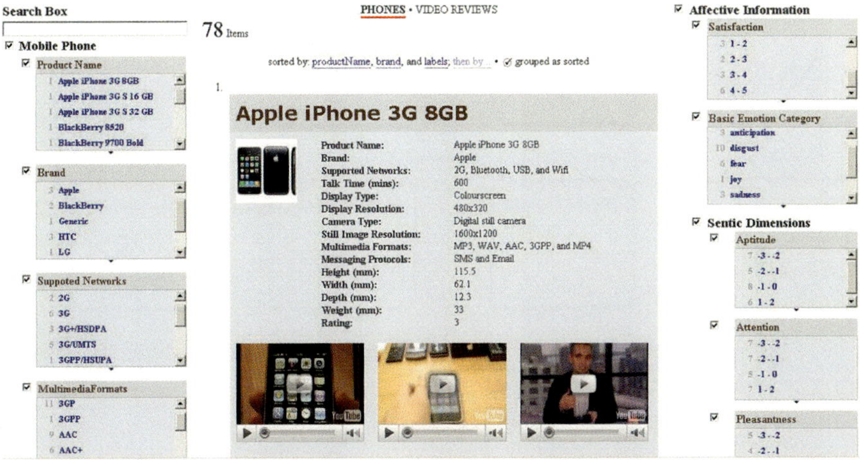

Fig. 4.4 A screenshot of the social media marketing tool. The faceted classification interface allows the user to navigate through both the explicit and implicit features of the different products (Source: [50])

gender, and nationality. Using such a tool a marketer can easily get an insight about the trend of a product, e.g., at the end of an advertising campaign, by observing how the number of reviews and the relative satisfaction evolve in time and by monitoring this trend for different campaign targets.

In order to evaluate the proposed system both on the level of opinion mining and sentiment analysis, its polarity detection accuracy was separately tested with a set of like/dislike-rated video reviews from YouTube and evaluated its affect recognition capabilities with a corpus of mood-tagged blogs from LiveJournal. In order to evaluate the system in terms of polarity detection accuracy, YouTube Data API was exploited to retrieve from YouTube database the ratings relative to the 220 video reviews previously selected for displaying in the faceted classification interface. On YouTube, in fact, users can express their opinions about videos either by adding comments or by simply rating them using a like/dislike button. YouTube Data API makes this kind of information available by providing, for each video, number of raters and average rating, i.e., sum of likes and dislikes divided by number of raters.

This information is expressed as a float $\in [1, 5]$ and indicates if a video is generally considered as bad (float $\in [1, 3]$) or good (float $\in [3, 5]$). This information was compared with the polarity values previously extracted by employing sentic computing on the comments relative to each of the 220 videos. True positives were identified as videos with both an average rating $\in [3, 5]$ and a polarity $\in [0, 1]$ (for positively rated videos), or videos with both an average rating $\in [1, 3]$ and a polarity $\in [-1, 0]$ (for negatively rated videos). The evaluation showed that, by using the system to perform polarity detection, negatively and positively rated videos (37.7 % and 62.3 % of the total respectively) can be identified with precision of 97.1 % and recall of 86.3 % (91.3 % F-measure).

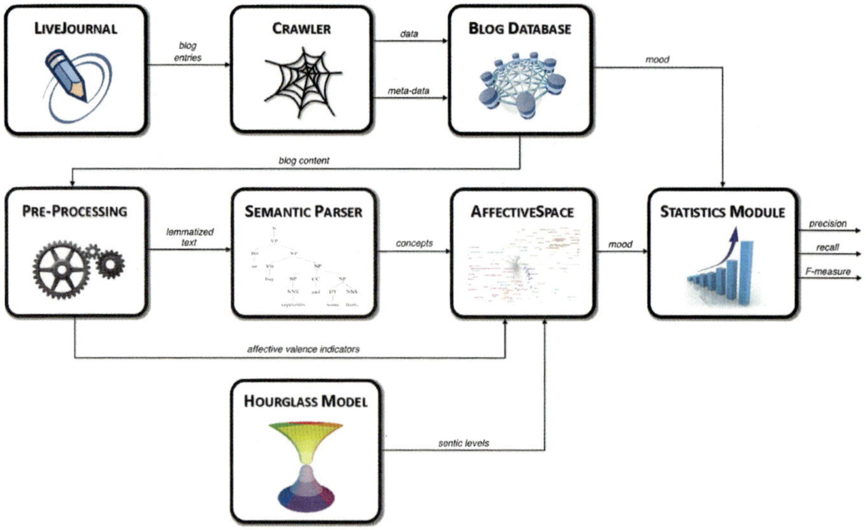

Fig. 4.5 Sentics extraction evaluation. The process extracts sentics from posts in the LiveJournal database, and then compare inferred emotional labels with the relative mood tags in the database (Source: [50])

Since no mood-labeled dataset about commercial products is currently available, the LiveJournal database was used to test the system's affect recognition capabilities. For this test, a reduced set of 10 moods has been considered, specifically, 'ecstatic', 'happy', 'pensive', 'surprised', 'enraged', 'sad', 'angry', 'annoyed', 'scared', and 'bored'. All LiveJournal accounts have Atom, RSS, and other data feeds that show recent public entries, friend relationships, and interests. Unfortunately, it is not possible to get mood-tagged blogposts via data feeds, hence, an ad hoc crawler had to be designed. After retrieving and storing relevant data and metadata for a total of 5,000 posts, the sentics extraction process was conducted on each of these and outputs were compared with the relative LiveJournal mood tags, in order to compute recall and precision rates as evaluation metrics (Fig. 4.5).

On average, each post contained around 140 words and, from it, about 4 affective valence indicators and 60 sentic vectors were extracted. According to this information, mood-labels were assigned to each post and compared with the corresponding LiveJournal mood tags, obtaining very good accuracy for each of the 10 selected moods (Table 4.2). Among these, 'happy' and 'sad' posts were identified with particularly high precision (89.2 % and 81.8 %, respectively) and decorous recall rates (76.5 % and 68.4 %). The F-measure values obtained, hence, were significantly good (82.3 % and 74.5 %, respectively), especially if compared to the corresponding F-measure rates of a standard keyword spotting system based on a set of 500 affect words (65.7 % and 58.6 %).

Table 4.2 Evaluation results of the sentics extraction process. Precision, recall, and F-measure rates are calculated for ten different moods by comparing the engine output with LiveJournal mood tags (Source: [50])

Mood	Precision (%)	Recall (%)	F-measure (%)
Ecstatic	73.1	61.3	66.6
Happy	**89.2**	**76.5**	**82.3**
Pensive	69.6	52.9	60.1
Surprised	81.2	65.8	72.6
Enraged	68.9	51.6	59.0
Sad	**81.8**	**68.4**	**74.5**
Angry	81.4	53.3	64.4
Annoyed	77.3	58.7	66.7
Scared	82.6	63.5	71.8
Bored	70.3	55.1	61.7

4.1.3 Sentic Album

Efficient access to online personal pictures requires the ability to properly annotate, organize, and retrieve the information associated with them. While the technology to search personal documents has been available for some time, the technology to manage personal images is much more challenging. This is mainly due to the fact that, even if images can be roughly interpreted automatically, many salient features exist only in the user's mind. The only way for a system to accordingly index personal images, hence, is to try to capture and process such features.

Existing content based image retrieval (CBIR) systems such as QBIC [126], Virage [18], MARS [256], ImageGrouper [223], MediAssist [228], CIVR [284], EGO [315], ACQUINE [101], and K-DIME [28] have attempted to build IUIs capable of retrieving pictures according to their intrinsic content through statistics, pattern recognition, signal processing, computer vision, SVM, and ANN.

All such techniques, however, appeared too weak to bridge the gap between the data representation and the images' conceptual models in the user's mind. Image meta search engines such as Webseek [290], Webseer [128], PicASHOW [185], IGroup [159], or Google,[8] Yahoo,[9] and Bing[10] Images, on the other hand, rely on tags associated with online pictures but, in the case of personal photo management, users are unlikely to expend substantial effort to manually classify and categorise images in the hopes of facilitating future retrieval. Moreover these techniques, as they depend on keyword-based algorithms, often miss potential connections between words expressed through different vocabularies or concepts that exhibit implicit semantic connectedness. In order to properly deal with photo metadata and, hence, effectively annotate images, in fact, it is necessary to work at a semantic, rather than syntactic, level.

---

[8]http://google.com/images

[9]http://images.search.yahoo.com

[10]http://bing.com/images

A good effort in this sense has been made within the development of ARIA [190], a software agent which aims to facilitate the storytelling task by opportunistically suggesting photos that may be relevant to what the user is typing. ARIA goes beyond the naïve approach of suggesting photos by simply matching keywords in a photo annotation with keywords in the story, as it also takes into account semantically related concepts. A similar approach has been followed by Raconteur [78], a system for conversational storytelling that encourages people to make coherent points, by instantiating large-scale story patterns and suggesting illustrative media. It exploits a large common-sense knowledge base to perform NLP in real-time on a text chat between a storyteller and a viewer and recommends appropriate media items from a library. Both these approaches present a lot of advantages since concepts, unlike keywords, are not sensitive to morphological variation, abbreviations, or near synonyms. However, simply relying on a semantic knowledge base is not enough to infer the salient features that make different pictures more or less relevant in each user's mind.

To this end, Sentic Album [49] exploits AI and Semantic Web techniques to perform reasoning on different knowledge bases and, hence, infer both the cognitive and affective information associated with photo metadata. The system, moreover, supports this concept-level analysis with content and context based techniques, in order to capture all the different aspects of online pictures and, hence, provide users with an IUI that is navigable in real-time through a multi-faceted classification website. Much of what is called problem-solving intelligence, in fact, is really the ability to identify what is relevant and important in a context and to subsequently make that knowledge available just in time [191].

Cognitive and affective processes are tightly intertwined in everyday life [96]. The affective aspect of cognition and communication is recognized to be a crucial part of human intelligence and has been argued to be more fundamental in human behavior for ensuring success in social life than intellect [240, 318].

Emotions, in fact, influence our ability to perform common cognitive tasks, such as forming memories and communicating with other people. A psychological study, for example, showed that people asked to conceal emotional facial expressions in response to unpleasant and pleasant slides remembered the slides less well than control participants [32]. Similarly, a study of conversations revealed that romantic partners who were instructed to conceal both facial and vocal cues of emotion while talking about important relationship conflicts with each other, remembered less of what was said than did partners who received no suppression instructions [270]. Many studies have indicated that emotions both seem to improve memory for the gist of an event and to undermine memory for more peripheral aspects of the event [37, 84, 267, 324].

The idea, broadly, is that arousal causes a decrease in the range of cues an organism can take in. This narrowing of attention leads directly to the exclusion of peripheral cues, and this is why emotionality undermines memory for information at the event's edge. At the same time, this narrowing allows a concentration of mental resources on more central materials, and this leads to the beneficial effects of emotion on memory for the event's centre [177]. Hence, rather than assigning particular cognitive and affective valence to a specific visual stimulus, we more often balance the importance of personal pictures is according to how much information contained in them is pertinent to our lives, goals, and values (or perhaps, the lives and values of people we care about). For this reason, a bad quality picture can be ranked high in the mind of a particular user, if it reminds him/her of a notably important moment or person of his/her life.

Events, in fact, are likely to be organized in the human mind as interconnected concepts and most of the links relating such concepts are probably weighted by affect, as we tend to better recall memories associated with either very positive or very negative emotions, as well as we usually tend to more easily forget about concepts associated with very little or null affective valence. The problem, when trying to emulate such cognitive and affective processes, is that, while cognitive information is usually objective and unbiased, affective information is rather subjective and argumentative. For example, while in the cognitive domain 'car' is always a car and there is usually not much discussion about the correctness of retrieving an image showing a tree in an African savannah under the label 'landscape', there might be some discussion about whether the retrieved car is "cool" or just "nice", or whether the found landscape is "peaceful" or "dull" [139]. To this end, Sentic Album applies sentic computing techniques on pictures data and metadata to infer what really matters to each user in different online photos. In particular, the Annotation Module mainly exploits metadata such as descriptions, tags, and comments, termed 'conceptual metadata', associated with each image to extract its relative semantics and sentics and, hence, enhance the picture specification with its intrinsic cognitive and affective information. This concept-level annotation procedure is performed through the opinion-mining engine and it is supported with a parallel content- and context-level analysis.

The content based annotation, in particular, is performed through Python Imaging Library[11] (PIL), an external library for the Python[12] programming language that adds support for opening, manipulating, and saving many different image file formats. For every online personal picture, in particular, PIL is exploited to extract luminance and chrominance information and other image statistics, e.g., the total, mean, standard deviation, and variance of the pixel values.

The context-based annotation, in turn, exploits information such as timestamp, geolocation, and user interaction metadata. Such metadata, termed 'contextual metadata', are processed by the Context Deviser, a sub-module that extracts small bits of information suitable for storing in a relational database for re-use at a later time, i.e., time, date, city and country of caption, plus relevant user interaction metadata such as number and IDs of friends who viewed, commented, or liked the picture.

The concept-based annotation represents the core of the module as it allows Sentic Album to go beyond a mere syntactic analysis of the metadata associated with pictures. A big problem of manual image annotation, in fact, is the different vocabulary that different users (or even the same user) can use to describe the

[11] http://pythonware.com/products/pil

[12] http://python.org

content of a picture. The different expertise and purposes of tagging users, in fact, may result in tags that use various levels of abstraction to describe a resource: a photo can be tagged at the 'basic level' of abstraction [175] as 'cat', or at a superordinate level as 'animal', or at various subordinate levels below the basic level as 'Persian cat' or 'Felis silvestris catus longhair Persian'.

To overcome this problem, Sentic Album extends the set of available tags with related semantics and sentics and, to further expand the cognitive and affective metadata associated with each picture, it extracts additional common-sense and affective concepts from its description and comments. In particular, the conceptual metadata is processed by the opinion-mining engine (Fig. 4.6). The IsaCore sub-module, specifically, finds matches between the retrieved concepts and those previously calculated using CF-IOF and spectral association. CF-IOF weighting is exploited to find seed concepts for a set of a-priori categories, extracted from

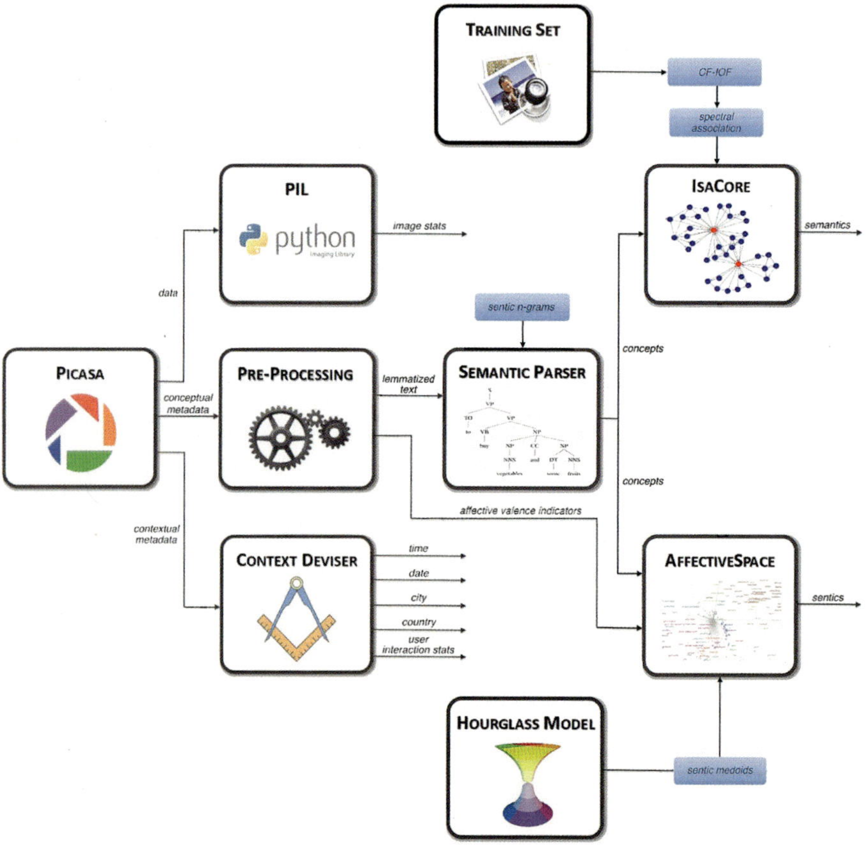

Fig. 4.6 Sentic Album's annotation module. Online personal pictures are annotated at three different levels: content level (PIL), concept level (opinion-mining engine) and context level (context deviser) (Source: [50])

Picasa[13] popular tags, meant to cover common topics in personal pictures, e.g., art, nature, friends, travel, wedding, or holiday. Spectral association is then used to expand this set with semantically related common-sense concepts. The retrieved concepts are also processed by the AffectiveSpace sub-module, which projects them into the vector space representation of AffectNet, clustered by means of sentic neurons, in order to infer the affective valence and the polarity associated with them.

Providing a satisfactory visual experience is one of the main goals for present-day electronic multimedia devices. All the enabling technologies for storage, transmission, compression, and rendering should preserve, and possibly enhance, image quality; and to do so, quality control mechanisms are required. Systems to automatically assess visual quality are generally known as objective quality metrics.

The design of objective quality metrics is a complex task because predictions must be consistent with human visual quality preferences. Human preferences are inherently quite variable and, by definition, subjective; moreover, in the field of visual quality, they stem from perceptual mechanisms that are not fully understood yet.

A common choice is to design metrics that replicate the functioning of the human visual system (HVS) to a certain extent, or at least that take into account its perceptual response to visual distortions by means of numerical features [166]. Although successful, these approaches come with a considerable computational cost, which makes them impractical for most real-time applications.

Computational intelligence paradigms allow for the handling of quality assessment from a different perspective, since they aim at mimicking quality perception instead of designing an explicit model of the HVS [196, 224, 266]. In the special case of personal pictures, perceived quality metrics can be computed not only at content level, but also at concept and context level. One of the primary reasons why people take pictures is to remember the emotions they felt on special occasions of their lives. Extracting and storing such affective information can be a key factor in improving future searches, as users seldom want to find photos matching general requirements. Users' criteria in browsing personal pictures, in fact, are more often related to the presence of a particular person in the picture and/or its perceived quality (e.g., to find a good photo of your mother). Satisfying this type of requirement is a tedious task as chronological ordering or classification by event does not help much. The process usually involves repeatedly trying to think of a matching picture and then looking for it. An exhaustive search (looking through the whole collection for all of the photos matching a requirement) would normally only be carried out in exceptional circumstances, such as following a death in the family. In order to accordingly rank personal photos, Sentic Album exploits data and metadata associated with them to extract useful information at content, concept, and context level and, hence, calculate the perceived quality of online pictures (PQOP):

[13] http://picasa.google.com

$$PQOP(p,u) = 3\frac{Content(p) * Concept(p,u) * Context(p,u)}{Content(p) + Concept(p,u) + Context(p,u)} \tag{4.3}$$

where *Content*, *Concept*, and *Context* (3Cs) are float $\in [0,1]$ representing image quality assessment values associated with picture p and user u, in terms of visual, conceptual, and contextual information, respectively. In particular, $Content(p)$ is computed from numerical features extracted through a reduced-reference framework for objective quality assessment based on extreme learning machine [105] and the color correlogram [155] of p; $Concept(p,u)$ specifies how much the picture p is relevant to the user u in terms of cognitive and affective information; finally, $Context(p,u)$ defines the degree of relevance of picture p for user u in terms of time, location, and user interaction. The 3Cs are all equally relevant for measuring how good a personal picture is to the eye of a user. According to the formula, in fact, if any of the 3Cs is null the PQOP is null as well, even though the remaining elements of the 3Cs have both maximum values, e.g., a perfect quality picture ($Content(p) = 1$) taken in the hometown of the user on the date of his birthday ($Context(p,u) = 1$), but depicting people he/she does not know and objects/places that are totally irrelevant for him/her ($Concept(p,u) = 0$).

The Storage Module is the middle-tier in which the outputs of the Annotation Module are stored, in a way that these can be easily accessible by the Search and Retrieval Module at a later time. The module stores information relative to photo data and metadata redundantly at three levels:

1. in a relational database fashion
2. in a Semantic Web format
3. in a matrix format

Sentic Album stores information in three main SQL databases (Fig. 4.7), that is a Content DB, for the information relative to data (image statistics), a Concept DB, for the information relative to conceptual metadata (semantics and sentics), and a Context DB, for the information relative to contextual metadata (timestamp, geolocation, and user interaction metadata). The Concept DB, in particular, consists of two databases, the Semantic DB and the Sentic DB, in which the cognitive and affective information associated with photo metadata are stored, respectively.

The Context DB, in turn, is divided into four databases: the Calendar, Geo, FOAF (Friend Of A Friend), and Interaction DBs, which contain the information relative to timestamp, geolocation, social links, and social interaction, respectively. These databases are also integrated with information coming from the web profile of the user such as user's DOB (for the Calendar DB), user's current location (for the Geo DB), or user's list of friends (for the FOAF DB). The FOAF DB, in particular, plays an important role within the Context DB since it provides the other peer databases with information relative to user's social connections, e.g., relatives' birthdays or friends' location. Moreover, the Context DB receives extra contextual information from the inferred semantics. Personal names in the conceptual metadata are recognized by building a dictionary of first names from the Web and combining them with regular expressions to recognize full names. These are added to the

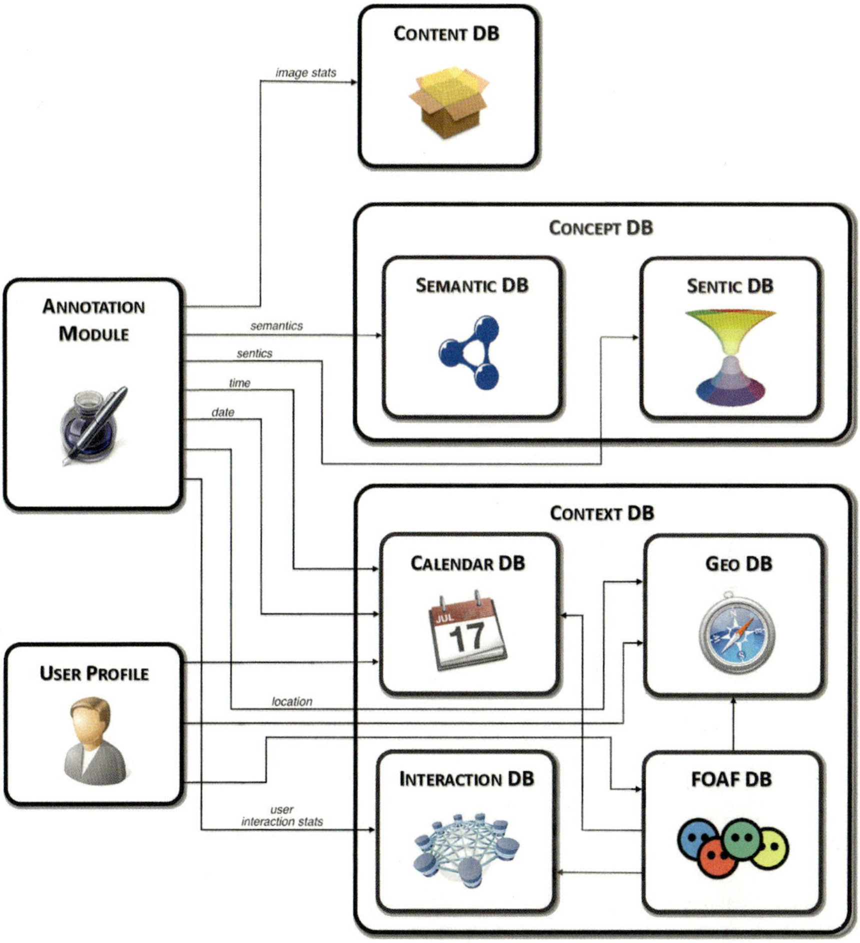

Fig. 4.7 Sentic Album's storage module. Image statistics are saved into the Content DB, semantics and sentics are stored into the Concept DB, timestamp and geolocation are saved into the Context DB (Source: [50])

database (in the FOAF DB) together with geographical places (in the Geo DB), which are also mined from databases on the Web and added to the parser's semantic lexicon.

As for the Semantic Web format [183], all the information related to pictures' metadata is stored in RDF/XML according to a set of predefined web ontologies. This operation aims to make the description of the semantics and sentics associated with pictures applicable to most online images coming from different sources, e.g., online photo sharing services, blogs, and social networks. To further this aim, it is necessary to standardize as much as possible the descriptors used in encoding the information about multimedia resources and people to which the images refer, in order to make it univocally interpretable and suitable to feed other applications. Hence, the ensemble of HEO, OMR, FOAF, and WNA is used again.

As for the storage of photo data and metadata in a matrix format, a dataset, termed '3CNet', is built, which integrates the information from the 3Cs in a unique knowledge base. The aim of this representation is to exploit principal component analysis (PCA) to later organize online personal images in a multi-dimensional vector space (as for AffectiveSpace) and, hence, reason on their similarity. 3CNet, in fact, is an $n \times m$ matrix whose rows are user's personal pictures IDs, whose columns are either content, concept, and context features (e.g., 'contains cold colors', 'conveys joy' or 'located in Italy'), and whose values indicate truth values of assertions. Therefore, in 3CNet, each image is represented by a vector in the space of possible features whose values are $+1$, for features that produce an assertion of positive valence, -1, for features that produce an assertion of negative valence, and 0 when nothing is known about the assertion.

The degree of similarity between two images, then, is the dot product between their rows in 3CNet. The value of such a dot product increases whenever two images are described with the same feature and decreases when they are described by features that are negations of each other. The main aim of the Search and Retrieval Module is to provide users with an IUI that allows them to easily manage, search, and retrieve their personal pictures online (Fig. 4.8). Most of the existing photo management systems let users search for pictures through a keyword-based query, but results are hardly ever good enough since it is very difficult to come up with an ideal query from the user's initial request.

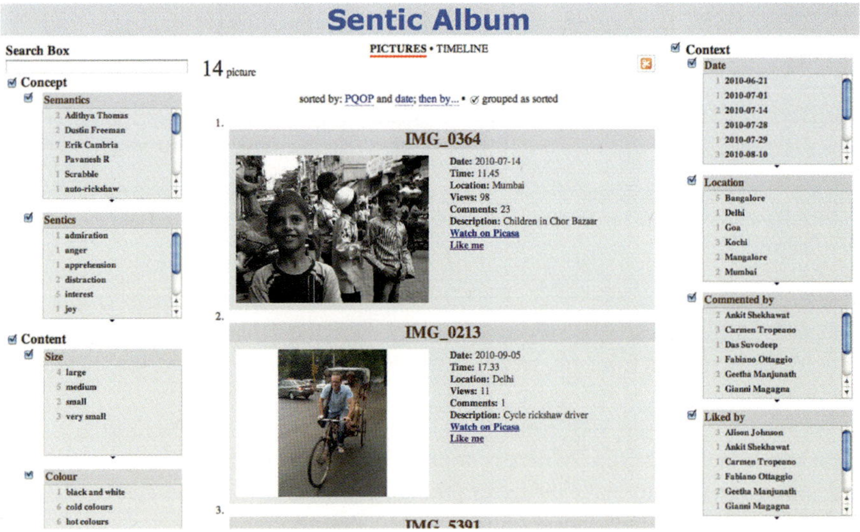

Fig. 4.8 Sentic Album's search and retrieval module. The IUI allows to browse personal images both by performing keyword-based queries and by adding/removing constraints on the facet properties (Source: [50])

The initial idea of an image the user has in mind before starting a search session, in fact, often deviates from the final results he/she will choose [316]. In order to let users start from a sketchy idea and then dynamically refine their search, the multi-faceted classification paradigm is adopted. Personal images are displayed in a dynamic gallery that can be ordered according to different parameters, either textual or numeric, that is visual features (e.g., color balance, hue, saturation, brightness, and contrast), semantics (i.e., common-sense concepts such as `go_jogging` and `birthday_party`, but also people and objects contained in the picture), sentics (i.e., emotions conveyed by the picture and its polarity) and contextual information (e.g., time of caption, location, and social information such as users who viewed/commented on the picture).

In particular, NLP techniques similar to those used to process the image conceptual metadata are employed to analyze the text typed in the search box and, hence, perform queries on the SQL databases of the Storage Module. The order of visualization of the retrieved images is given by the PQOP, so that images containing more relevant information at content, concept, and context level are first displayed. If, for example, the user is looking for pictures of his/her partner, Sentic Album initially proposes photos representing important events such as first date, first childbirth or honeymoon, that is, pictures with high PQOP. Storage Module's 3CNet is also exploited in the IUI, in order to find similar pictures.

Towards the end of a search, the user sometimes may be interested in finding pictures similar to one of those so far obtained, even if this does not fulfill the constraints currently set via the facets. To serve this purpose, every picture is provided with a 'like me' button that opens a new Exhibit window displaying content, concept, and context related images, independently of any constraint. Picture similarity is calculated by means of PCA and, in particular, through TSVD, as for AffectiveSpace. The number of singular values to be discarded (in order to reduce the dimensionality of 3CNet and, hence, reason on picture similarity) is chosen according to the total number of user's online personal pictures and the amount of available metadata associated with them, i.e., according to size and density of 3CNet. Thus, by exploiting the information sharing property of TSVD, images specified by similar content, concept, and context are likely to have similar features and, hence, tend to fall near each other in the built-in vector space. Finally, the IUI also offers to display images according to date of caption on a timeline. Chronology, in fact, is a key categorization concept for the management of personal pictures. Having the collection in chronological order is helpful for locating particular photos or events, since it is usually easier to remember when an event occurred relative to other events, as opposed to remembering its absolute date and time [179].

Many works dealing with object detection, scene categorization, or content analysis on the cognitive level have been published, trying to bridge the semantic gap between represented objects and high-level concepts associated with them [187]. However, where affective retrieval and classification of digital media is concerned, publications, and especially benchmarks, are very few [199]. To overcome the lack of availability of relevant datasets, the performance and the user-friendliness of

Table 4.3 Assessment of Sentic Album's accuracy in inferring the cognitive (topic tags) and affective (mood tags) information associated with the conceptual metadata typical of personal photos (Source: [50])

LiveJournal Tag	Precision (%)	Recall (%)	F-measure (%)
Art	62.9	55.6	59.0
Friends	**77.2**	**65.4**	**70.8**
Wedding	71.3	60.4	65.4
Holiday	68.9	59.2	63.7
Travel	**81.6**	**71.1**	**75.9**
Nature	67.5	61.8	64.5

Sentic Album were tested on a topic and mood tagged evaluation dataset and through a usability test on a pool of 18 Picasa regular users, respectively.

For the system performance testing, in particular, 1,000 LiveJournal posts with labels matching Picasa tags such as 'friends', 'travel', and 'holiday', were selected in order to collect natural language text that is likely to have the same semantics as the conceptual metadata typical of personal photos. The classification test, hence, concurrently estimated the capacity of the system to infer both the cognitive and affective information (topic and mood tags, respectively) usually associated with online personal pictures (Table 4.3).

For the usability test, users were asked to freely browse their online personal collections using Sentic Album IUI and to retrieve particular sets of pictures, in order to judge both usability and accuracy of the interface. Common queries included "find a funny picture of your best friend", "search for the shots of your last summer holiday", "retrieve pictures of you with animals", "find an image taken on Christmas 2009", "search for pictures of you laughing", and "find a good picture of your mom". From the test, it emerged that users really appreciate being able to dynamically and quickly set/remove constraints in order to display specific batches of pictures (which they cannot do in Picasa).

After the test session, participants were asked to fill-in an online questionnaire in which they were asked to rate, on a five-level scale, each single functionality of the interface according to their perceived utility. Concept facets and timeline, in particular, were found to be the most used by participants for search and retrieval tasks (Table 4.4). Users also really appreciated the 'like me' functionality, which was generally able to propose very relevant (semantically and affectively related) pictures (again not available in Picasa). When freely browsing their collections, users were particularly amused by the ability to navigate their personal pictures according to the emotion these conveyed, even though they did not always agree with the results.

Additionally, participants were not very happy with the accuracy of the search box, especially if they searched for one particular photo out of the entire collection. However, they always very much appreciated the order in which the pictures were proposed, which allowed them to quickly have all the most relevant pictures available as first results. 83.3 % of test users declared that, despite not being as nifty

Table 4.4 Perceived utility of the different interface features by 18 Picasa regular users. Participants particularly appreciated the usefulness of concept facets and timeline, for search and retrieval tasks (Source: [50])

Feature	Not at all (%)	Just a little (%)	Somewhat (%)	Quite a lot (%)	Very much (%)
Concept facets	0	0	5.6	5.6	88.8
Content facets	77.8	16.6	5.6	0	0
Context facets	16.6	11.2	5.6	33.3	33.3
Search box	0	11.2	16.6	33.3	38.9
Like me	0	5.6	5.6	16.6	72.2
Timeline	0	0	0	16.6	83.4
Sorting	11.2	33.3	33.3	16.6	5.6

as Picasa, Sentic Album is a very good photo management tool (especially for its novel semantic faceted search and PQOP functionalities) and they hope they could still be using it because, in the end, what really counts when browsing personal pictures is to find good matches in the shortest amount of time.

4.2 Development of HCI Systems

Human computer intelligent interaction is an emerging field aimed at providing natural ways for humans to use computers as aids. It is argued that, for a computer to be able to interact with humans, it needs to have the communication skills of humans. One of these skills is the affective aspect of communication, which is recognized to be a crucial part of human intelligence and has been argued to be more fundamental in human behavior and success in social life than intellect [240, 318]. Emotions influence cognition and, therefore, intelligence, especially when it involves social decision-making and interaction.

The latest scientific findings indicate that emotions play an essential role in decision-making, perception, learning, and more. Most of the past research on affect sensing has considered each sense such as vision, hearing, and touch in isolation. However, natural human-human interaction is multi-modal: we communicate through speech and use body language (posture, facial expressions, gaze) to express emotion, mood, attitude, and attention. To this end, a novel fusion methodology is proposed, which is able to fuse any number of unimodal categorical modules, with very different time-scales, output labels, and recognition success rates, in a simple and scalable way. In particular, such a methodology is exploited to fuse the outputs of the opinion-mining engine with the ones of a facial expression analyzer (Sect. 4.2.1). This section, moreover, illustrates how the engine can be exploited for the development of HCI applications in fields such as instant messaging (IM) (Sect. 4.2.2) and multimedia management (Sect. 4.2.3).

4.2.1 Sentic Blending

Subjectivity and sentiment analysis are the automatic identification of private states of the human mind (i.e., opinions, emotions, sentiments, behaviors and beliefs). Further, subjectivity detection focuses on identifying whether data is subjective or objective. Wherein, sentiment analysis classifies data into positive, negative and neutral categories and, hence, determines the sentiment polarity of the data.

To date, most of the works in sentiment analysis have been carried out on natural language processing. Available dataset and resources for sentiment analysis are restricted to text-based sentiment analysis only. With the advent of social media, people are now extensively using the social media platform to express their opinions. People are increasingly making use of videos (e.g., YouTube, Vimeo, VideoLectures), images (e.g., Flickr, Picasa, Facebook) and audios (e.g., podcasts) to air their opinions on social media platforms. Thus, it is highly crucial to mine opinions and identify sentiments from the diverse modalities. So far the field of multi-modal sentiment analysis has not received much attention [217], and no prior work has specifically addressed extraction of features and fusion of information extracted from different modalities.

Here, we discuss extraction process from different modalities is discussed, as well as the way these are exploited to build a novel multi-modal sentiment analysis framework. For the experiment, datasets from YouTube originally developed by [217] were used. Several supervised machine-learning-based classifiers were employed for the sentiment classification task. The best performance has been obtained with ELM. Research in this field is rapidly growing and attracting the attention of both academia and industry alike. This combined with advances in signal processing and AI has led to the development of advanced intelligent systems that intend to detect and process affective information contained in multi-modal sources. The majority of such state-of-the-art frameworks however, rely on processing a single modality, i.e., text, audio, or video. Further, all of these systems are known to exhibit limitations in terms of meeting robustness, accuracy, and overall performance requirements, which, in turn, greatly restrict the usefulness of such systems in real-world applications.

The aim of multi-sensor data fusion is to increase the accuracy and reliability of estimates [262]. Many applications, e.g., navigation tools, have already demonstrated the potential of data fusion. This depicts the importance and feasibility of developing a multi-modal framework that could cope with all three sensing modalities: text, audio, and video in human-centric environments. The way humans communicate and express their emotions and sentiments can be expressed as multi-modal. The textual, audio, and visual modalities are concurrently and cognitively exploited to enable effective extraction of the semantic and affective information conveyed during communication.

With significant increase in the popularity of social media like Facebook and YouTube, many users tend to upload their opinions on products in video format. On the contrary, people wanting to buy the same product, browse through on-line reviews and make their decisions. Hence, the market is more interested in mining opinions from video data rather than text data. Video data may contain more cues to identify sentiments of the opinion holder relating to the product. Audio data within a video expresses the tone of the speaker, and visual data conveys the facial expressions, which in turn help to understand the affective state of the users. The video data can be a good source for sentiment analysis but there are major challenges that need to be overcome. For example, expressiveness of opinions vary from person to person [217]. A person may express his or her opinions more vocally while others may express them more visually.

Hence, when a person expresses his opinions with more vocal modulation, the audio data may contain most of the clues for opinion mining. However, when a person is communicative through facial expressions, then most of the data required for opinion mining, would have been found in facial expressions. So, a generic model needs to be developed which can adapt itself for any user and can give a consistent result. The proposed multi-modal sentiment classification model is trained on robust data, and the data contains the opinions of many users. Here, we show that the ensemble application of feature extraction from different types of data and modalities enhances the performance of our proposed multi-modal sentiment system.

Sentiment analysis and emotion analysis both represent the private state of the mind and to-date, there are only two well known state-of-the-art methods [217] in multi-modal sentiment analysis. Next, the research done so far in both sentiment and emotion detection using visual and textual modality is described. Both feature extraction and feature fusion are crucial for the development of a multi-modal sentiment-analysis system. Existing research on multi-modal sentiment analysis can be categorized into two broad categories: those devoted to feature extraction from each individual modality, and those developing techniques for the fusion of features coming from different modalities.

4.2.1.1 Video: Emotion and Sentiment Analysis from Facial Expressions

In 1970, Ekman et al. [114] carried out extensive studies on facial expressions. Their research showed that universal facial expressions provide sufficient clues to detect emotions. They used anger, sadness, surprise, fear, disgust, and joy as six basic emotion classes. Such basic affective categories are sufficient to describe most of the emotions expressed by facial expressions. However, this list does not include the emotion expressed through facial expression by a person when he or she shows disrespect to someone; thus, a seventh basic emotion, contempt, was introduced by Matsumoto [205]. Ekman et al. [116] developed a facial expression coding system (FACS) to code facial expressions by deconstructing a facial expression into a set of action units (AU). AUs are defined via specific facial muscle movements. An AU consists of three basic parts: AU number, FACS name, and muscular basis. For example, for AU number 1, the FACS name is inner brow raiser and it is explicated via frontalis pars medialis muscle movements. In consideration to emotions, Friesen and Ekman [130] proposed the emotional facial action coding system (EFACS). EFACS defines the sets of AUs that participate in the construction of facial expressions expressing specific emotions.

The Active Appearance Model [99, 178] and Optical Flow-based techniques [167] are common approaches that use FACS to understand expressed facial expressions. Exploiting AUs as features like k nearest neighbors (KNN), Bayesian networks, hidden Markov models (HMM), and ANNs [314] has helped many researchers to infer emotions from facial expressions. All such systems, however, use different, manually crafted corpora, which makes it impossible to perform a comparative evaluation of their performance.

4.2.1.2 Audio: Emotion and Sentiment Recognition from Speech

Recent studies on speech-based emotion analysis [91, 99, 106, 160, 222] have focused on identifying several acoustic features such as fundamental frequency (pitch), intensity of utterance [76], bandwidth, and duration.

The speaker-dependent approach gives much better results than the speaker-independent approach, as shown by the excellent results of Navas et al. [225], where about 98 % accuracy was achieved by using the Gaussian mixture model (GMM) as a classifier, with prosodic, voice quality as well as Mel frequency cepstral coefficients (MFCC) employed as speech features. However, the speaker-dependent approach is not feasible in many applications that deal with a very large number of possible users (speakers).

For speaker-independent applications, the best classification accuracy achieved so far is 81 % [16], obtained on the Berlin Database of Emotional Speech (BDES) [38] using a two-step classification approach and a unique set of spectral, prosodic, and voice features, selected with the Sequential Floating Forward Selection (SFFS) algorithm [259]. As per the analysis of Scherer et al. [282], the human ability to recognize emotions from speech audio is about 60 %. Their study shows that sadness and anger are detected more easily from speech, while the recognition of joy and fear is less reliable. Caridakis et al. [69] obtained 93.30 % and 76.67 % accuracy in identifying anger and sadness, respectively, from speech, using 377 features based on intensity, pitch, MFCC, Bark spectral bands, voiced segment characteristics, and pause length.

4.2.1.3 Text: Emotion and Sentiment Recognition from Textual Data

Affective content recognition in text is a rapidly developing area of natural language processing, which has garnered the attention of both research communities and industries in recent years. Sentiment analysis tools have numerous applications. For example, it helps companies to comprehend customer sentiments about products and, political parties to understand what voters feel about party's actions and proposals. Significant studies have been done to identify positive, negative, or neutral sentiment associated with words [312, 327], multi-words [61], phrases [329], sentences [273], and documents [235]. The task of automatically identifying fine grained emotions, such as anger, joy, surprise, fear, disgust, and sadness, explicitly or implicitly expressed in a text has been addressed by several researchers [12, 303]. So far, approaches to text-based emotion and sentiment detection rely mainly on rule-based techniques, bag of words modeling using a large sentiment or emotion lexicon [215], or statistical approaches that assume the availability of a large dataset annotated with polarity or emotion labels [334].

Several supervised and unsupervised classifiers have been built to recognize emotional content in texts [337]. The SNoW architecture [75] is one of the most useful frameworks for text-based emotion detection. In the last decade, researchers have been focusing on sentiment extraction from texts of different genres, such as news [121], blogs [193], Twitter messages [233], and customer reviews [149] to name a few. Sentiment extraction from social media helps to predict the popularity of a product release, results of election poll, etc. To accomplish this, several knowledge-based sentiment [121] and emotion [20] lexicons have been developed for word- and phrase-level sentiment and emotion analysis.

4.2.1.4 Studies on Multi-modal Fusion

The ability to perform multi-modal fusion is an important prerequisite to the successful implementation of agent-user interaction. One of the primary obstacles to multi-modal fusion is the development and specification of a methodology to integrate cognitive and affective information from different sources on different time scales and measurement values. There are two main fusion strategies: feature-level fusion and decision-level fusion.

Feature-level fusion [285] combines the characteristics extracted from each input channel in a 'joint vector' before any classification operations are performed. Some variations of such an approach exist, e.g., Mansoorizadeh et al. [203] proposed asynchronous feature-level fusion. Modality fusion at feature-level presents the problem of integrating highly disparate input features, suggesting that the problem of synchronizing multiple inputs while re-teaching the modality's classification system is a nontrivial task.

In decision-level fusion, each modality is modeled and classified independently. The unimodal results are combined at the end of the process by choosing suitable metrics, such as expert rules and simple operators including majority votes, sums, products and statistical weighting. A number of studies favor decision-level fusion as the preferred method of data fusion because errors from different classifiers tend to be uncorrelated and the methodology is feature-independent [341]. Bimodal fusion methods have been proposed in numerous instances [136, 260], but optimal information fusion configurations remain elusive.

4.2.1.5 Datasets Employed

The YouTube Dataset developed by [217] was used. Forty-seven videos were collected from the social media web site YouTube. Videos in the dataset were about different topics (for instance politics, electronics product reviews, etc.). The videos were found using the following keywords: opinion, review, product review, best perfume, toothpaste, war, job, business, cosmetics review, camera review, baby product review, I hate, I like [217]. The final video set had 20 female and 27 male speakers randomly selected from YouTube, with their age ranging approximately from 14 to 60 years. Although, they belonged to different ethnic backgrounds (e.g., Caucasian, African-American, Hispanic, Asian), all speakers expressed themselves in English.

The videos were converted to mp4 format with a standard size of 360×480. The length of the videos varied from 2 to 5 min. All videos were pre-processed to avoid the issues of introductory titles and multiple topics. Many videos on YouTube contained an introductory sequence where a title was shown, sometimes accompanied by a visual animation. To address this issue first 30 s was removed from each video. Morency et al. [217] provided transcriptions with the videos. Each video was segmented and each segment was labeled by a sentiment, thanks to [217]. Because of this annotation scheme of the dataset, textual data was available for this experiment.

The YouTube dataset was used in this experiment to build the multi-modal sentiment-analysis system, as well as to evaluate the system's performance (as shown later). SenticNet and EmoSenticNet [255] were also used. The latter is an extension of SenticNet containing about 13,741 common-sense knowledge concepts, including those concepts that exist in the WNA list, along with their affective labels in the set anger, joy, disgust, sadness, surprise, fear. In order to build a suitable knowledge base for emotive reasoning, ConceptNet and EmoSenticNet were merged through blending, a technique that performs inference over multiple sources of data simultaneously, taking advantage of the overlap between them.

It linearly combines two sparse matrices into a single matrix, in which the information between two initial sources is shared. Before performing blending, EmoSenticNet is represented as a directed graph similar to ConceptNet. For example, the concept `birthday_party` was assigned an emotion `joy`. These are considered as two nodes, and the assertion HasProperty is added on the edge directed from the node `birthday_party` to the node `joy`.

Next, the graphs were converted to sparse matrices in order to blend them. After blending the two matrices, TSVD was performed on the resulting matrix to discard the components that represented relatively small variations in the data. Only 100 significant components of the blended matrix were retained in order to produce a good approximation of the original matrix. The number 100 was selected empirically: the original matrix was found to be best approximated using 100 components.

4.2.1.6 Overview of the Experiment

First, an empirical method used for extracting the key features from visual and textual data for sentiment analysis is presented. Then, a fusion method employed to fuse the extracted features for automatically identifying the overall sentiment expressed by a video is described.

- In YouTube dataset each video was segmented into several parts. According to the frame rate of the video, each video segment is first converted into images. Then, for each video segment facial features are extracted from all images and the average is taken to compute the final feature vector. Similarly, the audio and textual features were also extracted from each segment of the audio signal and text transcription of the video clip, respectively.
- Next, the audio, visual and textual feature vectors are fused to form a final feature vector which contained the information of both audio, visual and textual data. Later, a supervised classifier was employed on the fused feature vector to identify the overall polarity of each segment of the video clip. On the other hand, an experiment on decision-level fusion was also carried out, which took the sentiment classification result from three individual modalities as inputs and produced the final sentiment label as an output.

Humans are known to express emotions in a number of ways, including, to a large extent, through the face. Facial expressions play a significant role in the identification of emotions in a multi-modal stream. A facial expression analyzer automatically identifies emotional clues associated with facial expressions, and classifies facial expressions in order to define sentiment categories and to discriminate between them. Positive, negative and neutral were used as sentiment classes in the classification problem. In the annotations provided with the YouTube dataset, each video was segmented into some parts and each of the sub segments was of few seconds duration. Every segment was annotated as either 1, 0, or -1 denoting positive, neutral and negative sentiment.

Using a MATLAB code, all videos in the dataset were converted to image frames. Subsequently, facial features from each image frame were extracted. To extract facial characteristic points (FCPs) from the images, the facial recognition software Luxand FSDK 1.7 was used. From each image, 66 FCPs were extracted; see examples in Table 4.5. The FCPs were used to construct facial features, which are defined as distances between FCPs; see examples in Table 4.6.

GAVAM [278] was also used to extract facial expression features from the face. Table 4.7 shows the extracted features from facial images. In this experiment, the features extracted by FSDK 1.7 were used along with the features extracted using GAVAM. If a segment of a video has n number of images, then the features from each image were extracted and the average of those feature values was taken in order to compute the final facial expression feature vector for a segment. An ELM classifier was used to build the sentiment analysis model from the facial expressions. Ten-fold cross validation was carried out on the dataset producing 68.60 % accuracy.

Table 4.5 Some relevant facial characteristic points (out of the 66 facial characteristic points detected by Luxand) (Source: [252])

Features	Description
0	Left eye
1	Right eye
24	Left eye inner corner
23	Left eye outer corner
38	Left eye lower line
35	Left eye upper line
29	Left eye left iris corner
30	Left eye right iris corner
25	Right eye inner corner
26	Right eye outer corner
41	Right eye lower line
40	Right eye upper line
33	Right eye left iris corner
34	Right eye right iris corner
13	Left eyebrow inner corner
16	Left eyebrow middle
12	Left eyebrow outer corner
14	Right eyebrow inner corner
17	Right eyebrow middle
54	Mouth top
55	Mouth bottom

Table 4.6 Some important facial features used for the experiment (Source: [252])

Features
Distance between right eye and left eye
Distance between the inner and outer corner of the left eye
Distance between the upper and lower line of the left eye
Distance between the left iris corner and right iris corner of the left eye
Distance between the inner and outer corner of the right eye
Distance between the upper and lower line of the right eye
Distance between the left iris corner and right iris corner of the right eye
Distance between the left eyebrow inner and outer corner
Distance between the right eyebrow inner and outer corner
Distance between top of the mouth and bottom of the mouth

Audio features were automatically extracted from each annotated segment of the videos. Audio features were also extracted using a 30 Hz frame-rate and a sliding window of 100 ms. To compute the features, the open source software OpenEAR [122] was used. Specifically, this toolkit automatically extracts pitch and voice intensity. Z-standardization was used to perform voice normalization.

The voice intensity was thresholded to identify samples with and without voice. Using openEAR 6373 features were extracted. These features includes several

Table 4.7 Features extracted using GAVAM from the facial features (Source: [252])

Features
The time of occurrence of the particular frame in milliseconds
The displacement of the face w.r.t X-axis. It is measured by the displacement of the normal to the frontal view of the face in the X-direction
The displacement of the face w.r.t Y-axis
The displacement of the face w.r.t Z-axis
The angular displacement of the face w.r.t X-axis. It is measured by the angular displacement of the normal to the frontal view of the face with the X-axis
The angular displacement of the face w.r.t Y-axis
The angular displacement of the face w.r.t Z-axis

statistical measures, e.g., max and min value, standard deviation, and variance, of some key feature groups. Some of the useful key features extracted by openEAR are described below.

- Mel frequency cepstral coefficients – MFCC were calculated based on the short time Fourier transform (STFT). First, log-amplitude of the magnitude spectrum was taken, followed by grouping and smoothing the fast Fourier transform (FFT) bins according to the perceptually motivated Mel-frequency scaling. The Jaudio tool provided the first 5 of 13 coefficients, which were found to produce the best classification result.
- Spectral Centroid – Spectral Centroid is the center of gravity of the magnitude spectrum of the STFT. Here, $M_i[n]$ denotes the magnitude of the Fourier transform at frequency bin n and frame i. The centroid is used to measure the spectral shape. A higher value of the centroid indicates brighter textures with greater frequency. The spectral centroid is calculated as follows:

$$C_i = \frac{\sum_{i=0}^{n} nM_i[n]}{\sum_{i=0}^{n} M_i[n]}$$

- Spectral Flux – Spectral Flux is defined as the squared difference between the normalized magnitudes of successive windows: $F_i = \sum_{n=1}^{n} (N_t[n] - N_{t-1}[n])^2$ where $N_t[n]$ and $N_{t--1}[n]$ are the normalized magnitudes of the Fourier transform at the current frame t and the previous frame t–1, respectively. The spectral flux represents the amount of local spectral change.
- Beat histogram – It is a histogram showing the relative strength of different rhythmic periodicities in a signal, and is calculated as the auto-correlation of the RMS.
- Beat sum – This feature is measured as the sum of all entries in the beat histogram. It is a very good measure of the importance of regular beats in a signal.
- Strongest beat – It is defined as the strongest beat in a signal, in beats per minute and is found by finding the strongest bin in the beat histogram.

- Pause duration – Pause direction is the percentage of time the speaker is silent in the audio segment.
- Pitch – This is computed using the standard deviation of the pitch level for a spoken segment.
- Voice Quality – Harmonics to noise ratio in the audio signal.
- PLP – The Perceptual Linear Predictive Coefficients of the audio segment were calculated using the openEAR toolkit.

4.2.1.7 Fusion

Multi-modal fusion is the heart of any multi-modal sentiment analysis engine. As discussed before, there are two main fusion techniques: feature-level fusion and decision-level fusion. Feature-level fusion was implemented by concatenating the feature vectors of all three modalities, to form a single long feature vector.

This trivial method had the advantage of relative simplicity, yet was shown to produce significantly high accuracy. The feature vectors of each modality were concatenated into a single feature vector stream. This feature vector was then used for classifying each video segment into sentiment classes. To estimate the accuracy, ten-fold cross validation was used.

In decision-level fusion, feature vectors were obtained from the above-mentioned methods but instead of concatenating the feature vectors as in feature-level fusion, a separate classifier for each modality was used. The output of each classifier was treated as a classification score. In particular, a probability score for each sentiment class was obtained from each classifier. In this case, as there are three sentiment classes, three probability scores were obtained from each modality. The final label of the classification was then calculated using a rule-based approach given below:

$$l' = \arg\max_i (q_1 s_i^a + q_2 s_i^v + q_3 s_i^t), i = 1, 2, 3, \ldots, C$$

where q_1, q_2 and q_3 represent weights for the three modalities. An equal-weighted scheme was adopted, so in this case $q_1 = q_2 = q_3 = 0.33$. C is the number of sentiment classes, and s_i^a, s_i^v and s_i^t denote the scores from audio, visual and textual modality, respectively.

4.2.1.8 Proof of Concept

A real-time multi-modal sentiment-analysis system [47, 250, 252] was developed based on the methods described above. The framework allows a user to express his or her opinions in front of a camera. Later, it splits the video into several parts where each segment is empirically set to 5 s duration and sentiment is extracted as explained above Fig. 4.9.

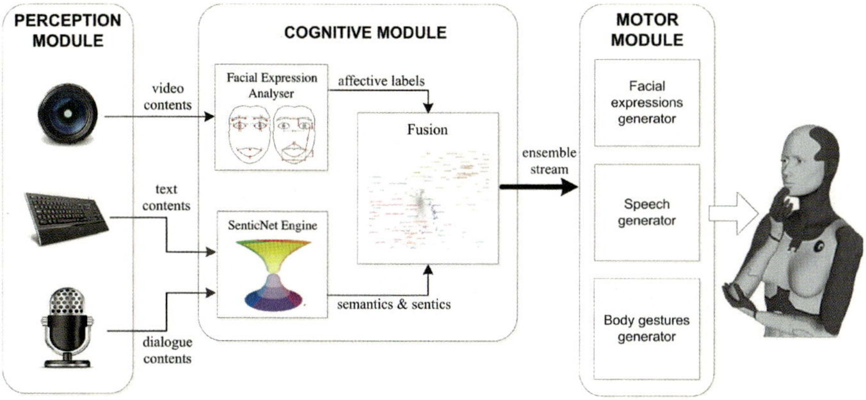

Fig. 4.9 Sentic blending framework (Source: [252])

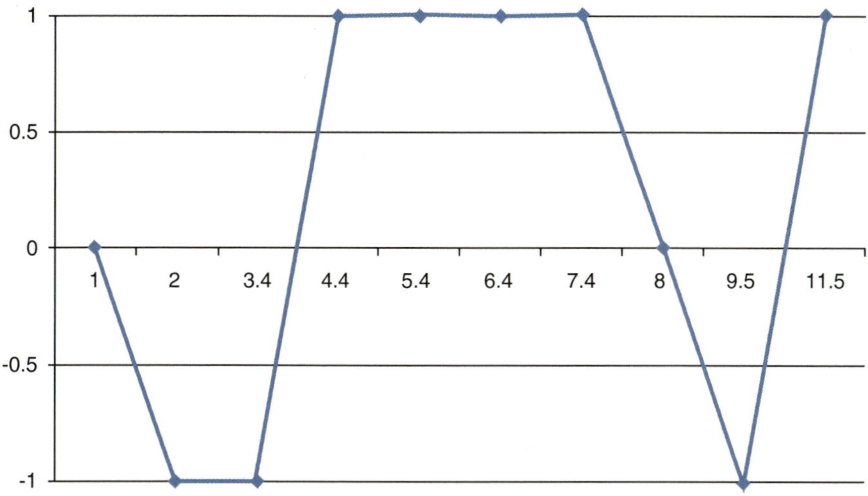

Fig. 4.10 Real-time multi-modal sentiment analysis of a YouTube product review video (Source: [252])

A transcriber was used to obtain the text transcription of the audio. Figure 4.10 shows that sentic blending analyzed a video and successfully detected its sentiment over time. The video related to a mobile and was collected from YouTube. Figure 4.10 shows the sentiment of the first 11.5 s of the video detected by the framework. In the initial 2 s, the reviewer expressed a positive sentiment about the product, followed by a negative sentiment from 2 to 4.4 s. This was followed by a positive review of the product expressed during the interval 4.4–8 s, and no sentiment expressed during the period 8–9.5 s. Finally, the reviewer expressed a positive sentiment about the product from 9.5 s till the end of the video.

Table 4.8 Results of feature-level fusion (Source: [252])

	Precision	Recall
Accuracy of the experiment carried out on textual modality	0.619	0.59
Accuracy of the experiment carried out on audio modality	0.652	0.671
Accuracy of the experiment carried out on video modality	0.681	0.676
Experiment using only visual and text-based features	0.7245	0.7185
Result obtained using visual and audio-based features	0.7321	0.7312
Result obtained using audio and text-based features	0.7115	0.7102
Accuracy of feature-level fusion of three modalities	0.782	0.771

Table 4.9 Results of decision-level fusion (Source: [252])

	Precision	Recall
Experiment using only visual and text-based features	0.683	0.6815
Result obtained using visual and audio-based features	0.7121	0.701
Result obtained using audio and text-based features	0.664	0.659
Accuracy of decision-level fusion of three modalities	0.752	0.734

Several supervised classifiers, namely Naïve Bayes, SVM, ELM, and Neural Networks, were employed on the fused feature vector to obtain the sentiment of each video segment. However, the best accuracy was obtained using the ELM classifier. Results for feature-level fusion are shown in Table 4.8, from which it can be seen that the proposed method outperforms [217] by 16.00 % in terms of accuracy.

Table 4.9 shows the experimental results of decision-level fusion. Tables 4.8 and 4.9 show the experimental results obtained when only *audio and text*, *visual and text*, *audio and visual* modalities were used for the experiment. It is clear from these tables, that the accuracy improves dramatically when audio, visual and textual modalities are used together. Finally, Table 4.8 also shows experimental results obtained when either the *visual* or *text* modality only, was used in the experiment.

The importance of each feature used in the classification task was also analyzed. The best accuracy was obtained when all features were used together. However, GAVAM features were found to be superior in comparison to the features extracted by Luxand FSDK 1.7.

Using only GAVAM features, an accuracy of 57.80 % was obtained for the visual features-based sentiment analysis task. However, for the same task, 55.64 % accuracy was obtained when only the features extracted by Luxand FSDK 1.7 were used. For the audio-based sentiment analysis task, MFCC and Spectral Centroid were found to produce a lower impact on the overall accuracy of the sentiment-analysis system. However, exclusion of those features led to a degradation of accuracy for the audio-based sentiment analysis task. The role of certain audio features like *time domain zero crossing*, *root mean square*, *compactness* was also experimentally evaluated, but no higher accuracy using any of them was obtained.

Table 4.10 Comparison of classifiers (Source: [252])

	Recall (%)	Training time
SVM	77.03	2.7 min
ELM	77.10	25 s
ANN	57.81	2.9 min

In the case of text-based sentiment analysis, concept-gram features were found to play a major role compared to SenticNet-based features. In particular, SenticNet-based features mainly helped detect associated sentiments in text in an unsupervised way. Note that the aim of sentic blending is to develop a multi-modal sentiment-analysis system where sentiment will be extracted from text in an unsupervised way using SenticNet as a knowledge base.

On the same training and test sets, the classification experiment was run using SVM, ANN and ELM. ELM outperformed ANN by 12 % in terms of accuracy (see Table 4.10). However, only a small difference in accuracy between the ELM and SVM classifiers was observed.

In terms of training time, the ELM outperformed SVM and ANN by a huge margin (Table 4.10). As the final goal is to develop a real-time multi-modal sentiment analysis engine, so ELM was preferred as a classifier as it provided the best performance in terms of both accuracy and training time.

4.2.2 Sentic Chat

Online communication is an extremely popular form of social interaction. Unlike face-to-face communication, online IM tools are extremely limited in conveying emotions or the context associated with a communication. Users have adapted to this environment by inventing their own vocabulary, e.g., by putting actions within asterisks ("I just came from a shower *shivering*"), or by using emoticons, by addressing a particular user in a group communication ("@Ravi"). Such evolving workarounds clearly indicate a latent need for a richer, more immersive user experience in social communication. This problem is addressed by exploiting the semantics and sentics associated with the on-going communication to develop an adaptive user interface (UI) capable to change according to content and context of the online chat. Popular approaches to enhance and personalize computer-mediated communication (CMC) include emoticons, skins, avatars, and customizable status messages.

However, all these approaches require explicit user configuration or action: the user needs to select the emoticon, status-message, or avatar that best represents him/her. Furthermore, most of these enhancements are static – once selected by the user, they do not adapt themselves automatically. There is some related work on automatically updating the status of the user by analyzing various sensor data available on mobile devices [211].

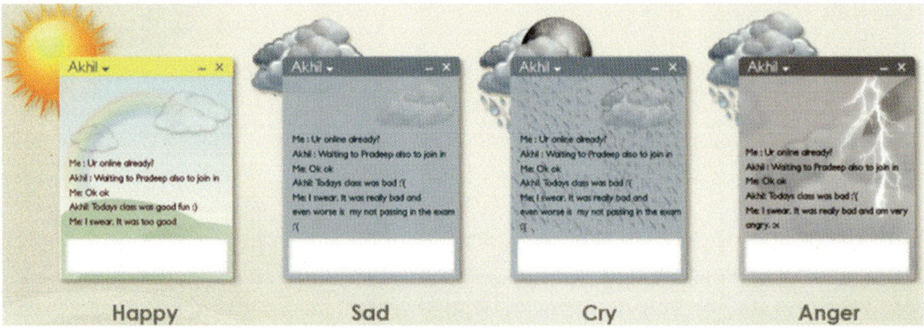

Fig. 4.11 A few screenshots of Sentic Chat IUI. Stage and actors gradually change, according to the semantics and sentics associated with the on-going conversation, to provide an immersive chat experience (Source: [50])

However, most of these personalization approaches are static and do not automatically adapt. The approach of Sentic Chat [73] is unique in that it is: intelligent, as it analyzes content and does not require explicit user configuration; adaptive, as the UI changes according to communication content and context; inclusive, as the emotions of one or more participants in the chat session are analyzed to let the UI adapt dynamically. The module architecture can be deployed either on the cloud (if the client has low processing capabilities) or on the client (if privacy is a concern). Most IM clients offer a very basic UI for text communication.

In Sentic Chat, the focus is on extracting the semantics and sentics embedded in the text of the chat session to provide an IUI that adapts itself to the mood of the communication. For this prototype application, the weather metaphor was selected, as it is scalable and has previously been used effectively to reflect the subject's mood [74] or content's 'flavour' [234]. In the proposed IUI, if the detected mood of the conversation is 'happy', the IUI will reflect a clear sunny day. Similarly a gloomy weather reflects a melancholy tone in the conversation (Fig. 4.11). Of course, this is a subjective metaphor – one that supposedly scales well with conversation analysis. In the future, other relevant scalable metaphors could be explored, e.g., colors [143].

The adaptive IUI primarily consists of three features: the stage, the actors, and the story. For any mapping, these elements pay a crucial role in conveying the feel and richness of the conversation mood, e.g., in the 'happy' conversation the weather 'clear sunny day' will be the stage, the actors will be lush green valley, the rainbow, and the cloud, which may appear or disappear as per the current conversation tone of the story. The idea is similar to a visual narrative of the mood the conversation is in; as the conversation goes on, the actors may come in or go off as per the tone of the thread. By analyzing the semantics and sentics associated with communication content (data) and context (metadata), the IUI may adapt to include images of landmarks from remote-user's location (e.g., Times Square), images about concepts in the conversation (pets, education, etc.), or time of day of remote user (e.g., sunrise or dusk).

The effectiveness of Sentic Chat was assessed through a usability test on a group of 6 regular chat users, who were asked to chat to each other pairwise for

Table 4.11 Perceived consistency with chat text of stage change and actor alternation. The evaluation was performed on a 130-min chat session operated by a pool of 6 regular chat users (Source: [50])

Feature	Not consistent (%)	Consistent (%)	Very consistent (%)
Stage change	0	83.3	16.7
Actor alternation	16.8	66.6	16.7

approximately 10 min (for a total of 130 min of chat data) and to rate the consistency with the story of both stage and actor alternation during the CMC (Table 4.11).

4.2.3 Sentic Corner

In a world in which web users are continuously blasted by ads and often compelled to deal with user-unfriendly interfaces, we sometimes feel like we want to evade the sensory overload of standard web pages and take refuge in a safe web corner, in which contents and design are in harmony with our current frame of mind. Sentic Corner [53] is an IUI that dynamically collects audio, video, images, and text related to the user's current feelings and activities as an interconnected knowledge base, which is browsable through a multi-faceted classification website. In the new realm of Web 2.0 applications, the analysis of emotions has undergone a large number of interpretations and visualizations, e.g., We Feel Fine,[14] MoodView,[15] MoodStats,[16] and MoodStream,[17] which have often led to the development of emotion-sensitive systems and applications.

Nonetheless, today web users still have to almost continuously deal with sensory-overloaded web pages, pop-up windows, annoying ads, user-unfriendly interfaces, etc. Moreover, even for websites uncontaminated by web spam, the affective content of the page is often totally unsynchronized with the user's emotional state. Web pages containing multimedia information inevitably carry more than just informative content. Behind every multimedia content, in fact, there is always an emotion.

Sentic Corner exploits this concept to build a sort of parallel cognitive/affective digital world in which the most relevant multimedia contents associated with the users' current moods and activities are collected, in order to enable them, whenever they want to evade from sensory-rich, overwrought, and earnest web pages, to take refuge in their own safe web corner. There is still no published study on the task

[14]http://wefeelfine.org

[15]http://moodviews.com

[16]http://moodstats.com

[17]http://moodstream.gettyimages.com

of automatically retrieving and displaying multimedia contents according to user's moods and activities, although the affective and semantic analysis of video, audio, and textual contents have been separately investigated extensively [139, 283, 299]. The most relevant commercial tool within this area is Moodstream, a mashup of several forms of media, designed to bring users music, images, and video according to the mood they manually select on the web interface.

Moodstream aims to create a sort of audio-visual ambient mix that can be dynamically modified by users by selecting from the presets of 'inspire', 'excite', 'refresh', 'intensify', 'stabilize', and 'simplify', e.g., mixtures of mood spectra on the Moodstream mixer such as happy/sad, calm/lively, or warm/cool. Users can start with a preset and then mix things up including the type of image transition, whether they want more or less vocals in their music selection, and how long images and video will stay, among other settings. In Moodstream, however, songs are not played entirely but blended into one another every 30 s and, even if the user has control on the multimedia flow through the mood presets, he/she cannot actually set a specific mood and/or activity as a core theme for the audio-visual ambient mix.

Sentic Corner, on the contrary, uses sentic computing to automatically extract semantics and sentics associated with user's status updates on micro-blogging websites and, hence, to retrieve relevant multimedia contents in harmony with his/her current emotions and motions. The module for the retrieval of semantically and affectively related music is termed Sentic Tuner. The relevant audio information is pulled from Stereomood,[18] an emotional online radio that provides music that best suits users' mood and activities. In the web interface, music is played randomly through an online music player with the possibility for the user to play, stop, and skip tracks. In Stereomood, music tracks are classified according to some tags that users are supposed to manually choose in order to access a list of semantically or affectively related songs. These tags are either mood tags (e.g., 'happy', 'calm', 'romantic', 'lonely', and 'reflective') or activity tags (such as 'reading', 'just woke up', 'dressing up', 'cleaning', and 'jogging'), the majority of which represent cognitive and affective knowledge contained in AffectiveSpace as common-sense concepts and emotional labels.

The Sentic Tuner uses the mood tags as centroids for affective blending and the activity tags as seeds for spectral association, in order to build a set of affectively and semantically related concepts respectively, which will be used at run-time to match the concepts extracted from user's micro-blogging activity. The Sentic Tuner also contains a few hundreds *rāgas* (Sanskrit for moods), which are melodic modes used in Indian classical music meant to be played in particular situations (mood, time of the year, time of the day, weather conditions, etc.). It is considered inappropriate to play *rāgas* at the wrong time (it would be like playing Christmas music in July, lullabies at breakfast, or sad songs at a wedding) so these are played just when semantics and sentics exactly match time and mood specifications in the *rāgas* database. Hence, once semantics and sentics are extracted from natural language

[18]http://stereomood.com

text through the opinion-mining engine, Stereomood API and the *rāgas* database are exploited to select the most relevant tracks to user's current feelings and activities.

Sentic TV is the module for the retrieval of semantically and affectively related videos. In particular, the module pulls information from Jinni,[19] a new site that allows users to search for video entertainment in many specific ways. The idea behind Jinni is to reflect how people really think and talk about what they watch.

It is based on an ontology developed by film professionals and new titles are indexed with an innovative NLP technology for analyzing metadata and reviews. In Jinni, users can choose from movies, TV shows, short films, and online videos to find specific genres or what they are in the mood to watch. In particular, users can browse videos by topic, mood, plot, genre, time/period, place, audience, and praise. Similarly to the Sentic Tuner, Sentic TV uses Jinni's mood tags as centroids for affective blending and the topic tags as seeds for spectral association, in order to retrieve affectively and semantically related concepts respectively. Time tags and location tags are also exploited in case relevant time-stamp and/or geo-location information is available within user's micro-blogging activity.

Sentic Corner also offers semantically and affectively related images through the Sentic Slideshow module. Pictures related to the user's current mood and activity are pulled from Fotosearch,[20] a provider of royalty free and rights managed stock photography that claims to be the biggest repository of images on the Web. Since Fotosearch does not offer a priori mood tags and activity tags, the CF-IOF technique is used on a set of 1000 manually tagged (according to mood and topic) tweets, in order to find seeds for spectral association (topic-tagged tweets) and centroids for affective blending (mood-tagged tweets). Each of the resulting concepts is used to retrieve mood and activity related images through the Fotosearch search engine.

The royalty free pictures, eventually, are saved in an internal database according to their mood and/or activity tag, in a way that they can be quickly retrieved at run-time, depending on user's current feelings and thoughts. The aim of Sentic Library is to provide book excerpts depending on user's current mood.

The module proposes random book passages users should read according to the mood they should be in while reading it and/or what mood they will be in when they have finished. The excerpt database is built according to '1001 Books for Every Mood: A Bibliophile's Guide to Unwinding, Misbehaving, Forgiving, Celebrating, Commiserating' [118], a guide in which the novelist Hallie Ephron serves up a literary feast for every emotional appetite. In the guide, books are labeled with mood tags such as 'for a good laugh', 'for a good cry', and 'for romance', but also some activity tags such as 'for a walk on the wild side' or 'to run away from home'.

As for Sentic TV and Sentic Tuner, Sentic Library uses these mood tags as centroids for affective blending and the topic tags as seeds for spectral association. The Corner Deviser exploits the semantic and sentic knowledge bases previously built by means of blending, CF-IOF and spectral association to find matches for the concepts extracted by the semantic parser and their relative affective information

[19] http://jinni.com

[20] http://fotosearch.com

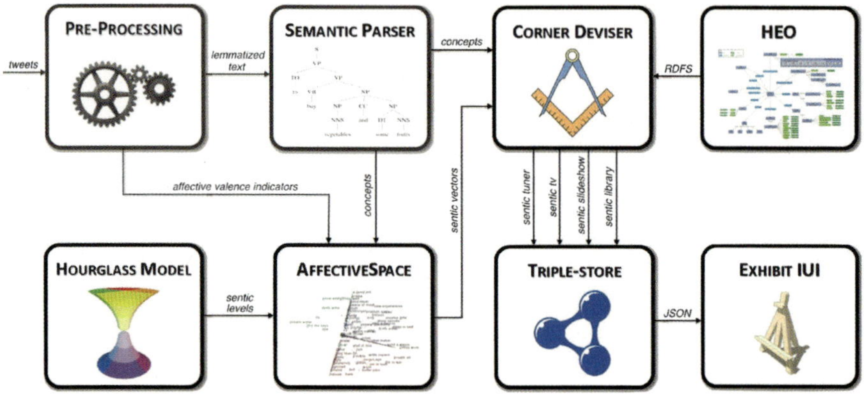

Fig. 4.12 Sentic Corner generation process. The semantics and sentics extracted from the user's micro-blogging activity are exploited to retrieve relevant audio, video, visual, and textual information (Source: [50])

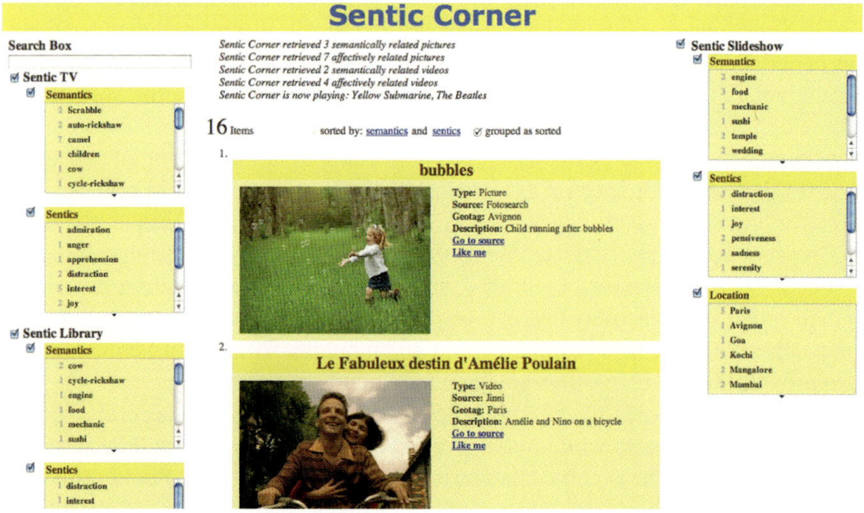

Fig. 4.13 Sentic Corner web interface. The multi-modal information obtained by means of Sentic Tuner, Sentic TV, Sentic Slideshow, and Sentic Library is encoded in RDF/XML for multi-faceted browsing (Source: [50])

inferred by AffectiveSpace. Such audio, video, visual, and textual information (namely Sentic Tuner, Sentic TV, Sentic Slideshow, and Sentic Library) is then encoded in RDF/XML according to HEO and stored in a triple-store (Fig. 4.12).

In case the sentics detected belong to the lower part of the Hourglass, the multimedia contents searched will have an affective valence opposite to the emotional charge detected, as Sentic Corner aims to restore the positive emotional equilibrium of the user, e.g., if the user is angry he/she might want to calm down.

The Exhibit IUI module, eventually, visualizes the contents of the Sesame database exploiting the multi-faceted categorization paradigm (Fig. 4.13). In order

Table 4.12 Relevance of audio, video, visual, and textual information assembled over 80 tweets. Because of their larger datasets, Sentic Tuner and Slideshow are the best-performing modules (Source: [50])

Content	Not at all (%)	Just a little (%)	Somewhat (%)	Quite a lot (%)	Very much (%)
Audio	0	11.1	11.1	44.5	33.3
Video	11.1	11.1	44.5	33.3	0
Visual	0	0	22.2	33.3	44.5
Textual	22.2	11.1	55.6	11.1	0

to test the relevance of multimedia content retrieval, an evaluation based on the judgements of eight regular Twitter users was performed. Specifically, users had to link Sentic Corner to their Twitter accounts and evaluate, over ten different tweets, how the IUI would react to their status change in terms of relevance of audio, video, visual, and textual information assembled by Sentic Corner. The multimedia contents retrieved turned out to be pretty relevant in most cases, especially for tweets concerning concrete entities and actions (Table 4.12).

4.3 Development of E-Health Systems

In health care, it has long been recognized that, although the health professional is the expert in diagnosing, offering help, and giving support in managing a clinical condition, the patient is the expert in living with that condition. Health-care providers need to be validated by someone outside the medical departments but, at the same time, inside the health-care system. The best candidate for this is not the doctor, the nurse, or the therapist, but the real end-user of health-care – none other than the patient him/herself.

Patient 2.0 is central to understanding the effectiveness and efficiency of services and how they can be improved. The patient is not just a consumer of the health-care system but a quality control manager – his/her opinions are not just reviews of a product/service but more like small donations of experience, digital gifts which, once given, can be shared, copied, moved around the world, and directed to just the right people who can use them to improve health-care locally, regionally, or nationally. Web 2.0 dropped the cost of voice, of finding others 'like me', of forming groups, of obtaining and republishing information, to zero. As a result, it becomes easy and rewarding for patients and carers to share their personal experiences with the health-care system and to research conditions and treatments.

To bridge the gap between this social information and the structured information supplied by health-care providers, the opinion-mining engine is exploited to extract the semantics and sentics associated with patient opinions over the Web.

In this way, the engine provides the real end-users of the health system with a common framework to compare, validate, and select their health-care providers (Sect. 4.3.1). This section, moreover, shows how the engine can be used as an embedded tool for improving patient reported outcome measures (PROMs) for health related quality of life (HRQoL), that is to record the level of each patient's physical and mental symptoms, limitations, and dependency (Sect. 4.3.2).

4.3.1 Crowd Validation

As Web 2.0 dramatically reduced the cost of communication, today it is easy and rewarding for patients and carers to share their personal experiences with the health-care system. This social information, however, is often stored in natural language text and, hence, intrinsically unstructured, which makes comparison with the structured information supplied by health-care providers very difficult. To bridge the gap between these data, which though different at structure level are similar at concept level, a patient opinion mining tool has been proposed to provide the end-users of the health system with a common framework to compare, validate, and select their health-care providers.

In order to give structure to online patient opinions, both the semantics and sentics associated with these are extracted in a way that it is possible to map them to the fixed structure of health-care data. This kind of data, in fact, usually consists of ratings that associate a polarity value to specific features of health-care providers such as communication, food, parking, service, staff, and timeliness. The polarity can either be a number in a fixed range or simply a flag (positive/negative). In the proposed approach, structure is added to unstructured data by building semantics and sentics on top of it (Fig. 4.14).

In particular, given a textual resource containing a set of opinions O about a set of topics T with different polarity $p \in [-1, 1]$, the subset of opinions $o \subseteq O$ is extracted, for each $t \in T$, and p is determined for each o. In other words, since each opinion can regard more than one topic and the polarity values associated with each topic are often independent from each other, a set of topics is extracted from each opinion and then, for each topic detected, the polarity associated with it is inferred. Once natural language data are converted to a structured format, each topic expressed in

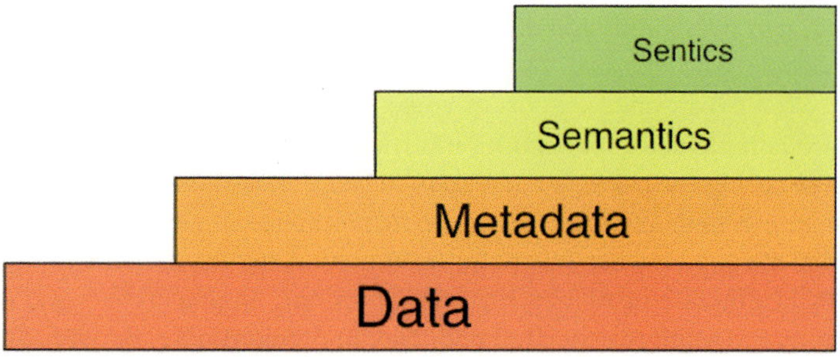

Fig. 4.14 The semantics and sentics stack. Semantics are built on the top of data and metadata. Sentics are built on the top of semantics, representing the affective information associated with these (Source: [50])

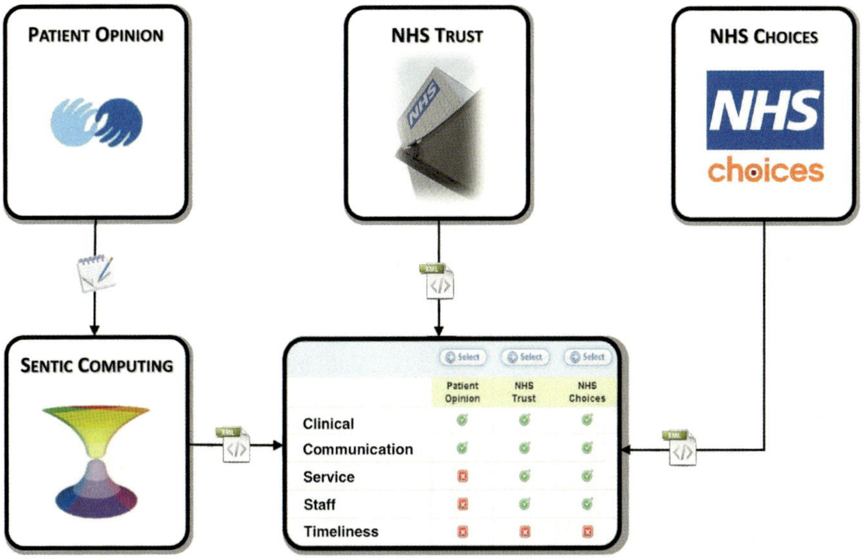

Fig. 4.15 The crowd validation schema. PatientOpinion stories are encoded in a machine-accessible format, in a way that they can be compared with the ratings provided by NHS choices and each NHS trust (Source: [50])

each patient opinion and its related polarity can be aggregated and compared. These can then be easily assimilated with structured health-care information contained in a database or available through an API.

This process is termed crowd validation [56] (Fig. 4.15), because of the feedback coming from the masses, and it fosters next-generation health-care systems, in which patient opinions are crucial in understanding the effectiveness and efficiency of health services and how they can be improved. Within this work, in particular, the opinion-mining engine is used to marshal PatientOpinion's social information in a machine-accessible and machine-processable format and, hence, compare it with the official hospital ratings provided by NHS Choices[21] and each NHS trust. The inferred ratings are used to validate the information declared by the relevant health-care providers (crawled separately from each NHS trust website) and the official NHS ranks (extracted using NHS Choices API). At the present time, crowd validation cannot be directly tested because of the impossibility to objectively assess the truthfulness of both patient opinions and official NHS ratings.

An experimental investigation has been performed over a set of 200 patient opinions about three different NHS trusts, for which self-assessed ratings were crawled from each hospital website and official NHS ranks were obtained through NHS Choices API. Results showed an average discrepancy of 39 % between official

[21] http://www.nhs.uk

and unofficial ratings, which sounds plausible as, according to Panorama,[22] 60 % of hospitals inspected in 2010 gave inaccurate information to the government in assessing their own performance.

4.3.2 Sentic PROMs

Public health measures such as better nutrition, greater access to medical care, improved sanitation, and more widespread immunization have produced a rapid decline in death rates across all age groups. Since there is no corresponding decline in birth rates, however, the average age of population is increasing exponentially. If we want health services to keep up with such monotonic growth, we need to automatize as much as possible the way patients access the health-care system, in order to improve both its service quality and timeliness. Everything we do that does not provide benefit to patients or their families, in fact, is a waste.

To this end, a new generation of short and easy-to-use tools to monitor patient outcomes and experience on a regular basis have been recently proposed by Benson et al. [26]. Such tools are quick, effective, and easy to understand, as they are very structured. However, they leave no space for those patients who would like to say something more. Patients, in fact, are usually keen on expressing their opinions and feelings in free text, especially if driven by particularly positive or negative emotions. They are often happy to share their health-care experiences for different reasons, e.g., because they seek for a sense of togetherness in adversity, because they benefited from others' opinions and want to give back to the community, for cathartic complaining, for supporting a service they really like, because it is a way to express themselves, because they think their opinions are important for others. When people have a strong feeling about a specific service they tried, they feel like expressing it. If they loved it, they want others to enjoy it. If they hated it, they want to warn others away.

Standard PROMs allow patients to easily and efficiently measure their HRQoL but, at the same time, they limit patients' capability and will to express their opinions about particular aspects of the health-care service that could be improved or important facets of their current health status. Sentic PROMs [42], in turn, exploit the ensemble application of standard PROMs and sentic computing to allow patients to evaluate their health status and experience in a semi-structured way, i.e., both through a fixed questionnaire and through free text.

PROMs provide a means of gaining an insight into the way patients perceive their health and the impact that treatments or adjustments to lifestyle have on their quality of life. Pioneered by Donabedian [108], health status research began during the late 1960s with works focusing on health-care evaluation and resource allocation. In particular, early works mainly aimed to valuate health states for policy and economic

[22]http://www.bbc.co.uk/programmes/b00rfqfm

evaluation of health-care programs, but devoted little attention to the practicalities of data collection [93, 123, 307]. Later works, in turn, aimed to develop lengthy health profiles to be completed by patients, leading to the term patient reported outcome [27, 321].

PROMs can provide a new category of real-time health information, which enables every level of the health service to focus on continuously improving those things that really matter to patients. The benefits of routine measurement of HRQoL include helping to screen for problems, promoting patient-centric care, aiding patients and doctors to take decisions, improving communication amongst multi-disciplinary teams, and monitoring progress of individual or groups of patients and the quality of care in a population. However, in spite of demonstrated benefits, routine HRQoL assessment in day-to-day practice remains rare as few patients are willing to spend the time needed to daily fill-in questionnaires, such as *SF-36* [323], *SF-12* [322], *Euroqol EQ-5D* [36], or the *Health Utilities Index* [146].

To overcome this problem, a new generic PROM, termed *howRu* [26], was recently proposed for recording the level of each patient's physical and mental symptoms, limitations, and dependency on four simple levels. The questionnaire was designed to take no more than a few seconds using electronic data collection and integration with electronic patient records as part of other routine tasks that patients have to do, such as booking appointments, checking in on arrival at clinic, and ordering or collecting repeat medication. The main aim of *howRu* is to use simple terms and descriptions to reduce the risk of ambiguity and to ensure that as many people as possible can use the measure reliably and consistently without training or support. The same approach has been employed to monitor also patient experience (*howRwe*) and staff satisfaction (*howRus*) on a regular basis. These questionnaires have been proved to be quick, effective, and easy to understand, as they are short, rigid, and structured. However, such structuredness can be very limiting, as it leaves no space to those patients who would like to say something more about their health or the service they are receiving. Patients, especially when driven by particularly positive or negative emotions, do want to express their opinions and feelings.

Sentic PROMs allow patients to assess their health status and health-care experience in a semi-structured way by enriching the functionalities of *howRu* with the possibility of adding free text (Fig. 4.16). This way, when patients are happy with simply filling-in the questionnaire, they can just leave the input text box blank but, when they feel like expressing their opinions and feelings, e.g., in the occasion of a particularly positive or negative situation or event, they can now do it in their own words. Hence, Sentic PROMs input data, although very similar at concept level, are on two completely different structure levels – structured (questionnaire selection) and unstructured (natural language).

As we would like to extract meaningful information from such data, the final aim of Sentic PROMs is to format the unstructured input and accordingly aggregate it with the structured data, in order to perform statistical analysis and pattern recognition. In particular, the gap between unstructured and structured data is bridged by means of sentic computing.

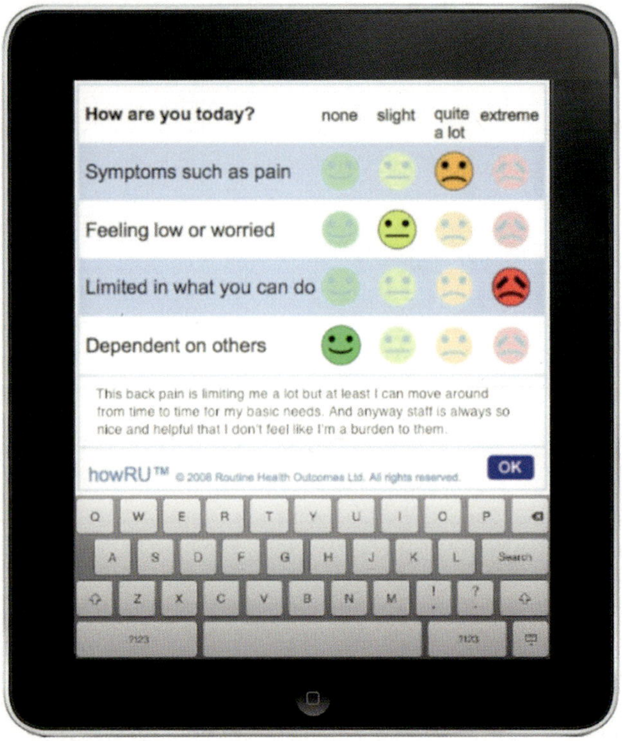

Fig. 4.16 Sentic PROMs prototype on iPad. The new interface allows patients to assess their health status and health-care experience both in a structured (questionnaire) and unstructured (free text) way (Source: [50])

Some of the benefits of structuring questionnaires include speed, effectiveness, and ease of use and understanding. However, such structuredness involves some drawbacks. A questionnaire, in fact, can limit the possibility of discovering new important patterns in the input data and can constrain users to omit important opinions that might be valuable for measuring service quality.

In the medical sphere, in particular, patients driven by very positive or very negative emotions are usually willing to detailedly express their point of view, which can be particularly valuable for assessing uncovered points, raising latent problems, or redesigning the questionnaire. To this end, Sentic PROMs adopt a semi-structured approach that allows patients to assess their health status and health-care experience both by filling in a four-level questionnaire and by adding free text. The analysis of free text, moreover, allows Sentic PROMs to measure patients' physio-emotional sensitivity. The importance of physio-emotional sensitivity in humans has been proven by recent health research, which has shown that individuals who feel loved and supported by friends and family, or even by a loving pet, tend to have higher survival rates following heart attacks than other cardiac patients who experience a sense of social isolation. Such concept is also reflected in natural language as we use terms such as ‘heartsick’, ‘broken-hearted’, and ‘heartache’ to describe extreme sadness and grief, idioms like ‘full of gall’ and ‘venting your spleen’ to describe anger, and expressions such as ‘gutless’, ‘yellow belly’, and ‘feeling kicked in the gut’ to describe shame. Human body contracts involuntarily when it feels emotional pain such as grief, fear, disapproval, shock, helplessness, shame, and terror, in the same reflex it does if physically injured. Such gripping reflex normally releases slowly, but if a painful experience is intense, or happens repeatedly, the physio-emotional grip does not release and constriction is retained in the body. Any repeated similar experience then layers on top of the original unreleased contraction, until we are living with layers of chronic tension, which constricts our bodies. The mind, in fact, may forget the origin of pain and tension, but the body does not.

In addition to HRQoL measurements, Sentic PROMs aim to monitor users’ physio-emotional sensitivity on a regular basis, as a means of patient affective modeling. In particular, the dimensional affective information coming from both questionnaire data (*howRu* aggregated score) and natural language data (sentic vectors) is stored separately by the system every time patients conclude a Sentic PROM session, and plotted on four different bi-dimensional diagrams. Such diagrams represent the pairwise fusion of the four dimensions of the Hourglass model and enable detection of more complex (compound) emotions that can be particularly relevant for monitoring patients’ physio-emotional sensitivity, e.g., frustration, anxiety, optimism, disapproval, and rejection.

A preliminary validation study was undertaken to examine the psychometric properties and construct validity of Sentic PROMs and to compare these with *SF-12*. In particular, 2,751 subjects with long-term conditions (average age 62, female 62.8 %), were classified by *howRu* score, primary condition, number of conditions suffered, age group, duration of illness, and area of residence. Across all six classifications, the correlation of the mean *howRu* scores with the mean values of the Physical Components Summary (*PCS-12*), the Mental Components Summary (*MCS-12*), and the sum of *PCS-12* + *MCS-12* were generally very high (0.91, 0.45, and 0.97, respectively).

Chapter 5
Conclusion

Human beings, viewed as behaving systems, are quite simple. The apparent complexity of our behavior over time is largely a reflection of the complexity of the environment in which we find ourselves.

Herbert Simon

Abstract The main aim of this book was to go beyond keyword-based approaches by further developing and applying common-sense computing and linguistic patterns to bridge the cognitive and affective gap between word-level natural language data and the concept-level opinions conveyed by these. This has been pursued through a variety of novel tools and techniques that have been tied together to develop an opinion-mining engine for the semantic analysis of natural language opinions and sentiments. The engine has then been used for the development of intelligent web applications in diverse fields such as Social Web, HCI, and e-health. This final section proposes a summary of contributions in terms of models, techniques, tools, and applications introduced by sentic computing, and lists some of its limitations.

Keywords Sentic models • Sentic tools • Sentic techniques • Sentic applications • Artificial intelligence

This chapter contains a summary of the contributions the book has introduced (Sect. 5.1), a discussion about limitations and future developments of these (Sect. 5.2).

5.1 Summary of Contributions

Despite significant progress, opinion mining and sentiment analysis are still finding their own voice as new inter-disciplinary fields. Engineers and computer scientists use machine learning techniques for automatic affect classification from video,

© Springer International Publishing Switzerland 2015

E. Cambria, A. Hussain, *Sentic Computing*, Socio-Affective Computing 1,
DOI 10.1007/978-3-319-23654-4_5

voice, text, and physiology. Psychologists use their long tradition of emotion research with their own discourse, models, and methods. This work has assumed that opinion mining and sentiment analysis are research fields inextricably bound to the affective sciences that attempt to understand human emotions. Simply put, the development of affect-sensitive systems cannot be divorced from the century-long psychological research on emotion. The emphasis on the multi-disciplinary landscape that is typical for emotion-sensitive applications and the need for common-sense sets this work apart from previous research on opinion mining and sentiment analysis.

In this book, a novel approach to opinion mining and sentiment analysis has been developed by exploiting both AI and linguistics. In particular, an ensemble of common-sense computing, linguistic patterns and machine learning has been employed for the sentiment analysis task of polarity detection. Such a framework has then been embedded in multiple systems in a range of diverse fields such as Social Web, HCI, and e-health. This section lists the models, techniques, tools, and applications developed within this work.

5.1.1 Models

1. **AffectNet**: an affective common-sense representation model built by integrating different kinds of knowledge coming from multiple sources;
2. **AffectiveSpace**: a vector space model built by means of semantic multi-dimensional scaling (MDS) for reasoning by analogy on affective common-sense knowledge;
3. **The Hourglass of Emotions**: a biologically-inspired and psychologically-motivated model for the representation and the analysis of human emotions.

5.1.2 Techniques

1. **Sentic Patterns**: linguistic rules that allow sentiment to flow from concept to concept based on the dependency relation of the input sentence and, hence, to generate a polarity value;
2. **Sentic Activation**: a bio-inspired two-level framework that exploits an ensemble application of dimensionality-reduction and graph-mining techniques;
3. **Sentic Blending**: scalable multi-modal fusion for the continuous interpretation of semantics and sentics in a multi-dimensional vector space;
4. **Crowd Validation**: a process for mining patient opinions and bridging the gap between unstructured and structured health-care data.

5.1.3 *Tools*

1. **SenticNet**: a semantic and affective resource that assigns semantics and sentics with 30,000 concepts (also accessible through an API and a Python package);
2. **Semantic Parser**: a set of semantic parsing techniques for effective multi-word commonsense expression extraction from unrestricted English text;
3. **Sentic Neurons**: ensemble application of multi-dimensional scaling and artificial neural networks for biologically-inspired opinion mining;
4. **GECKA**: a game engine for collecting common-sense knowledge from game designers through the development of serious games.

5.1.4 *Applications*

1. **Troll Filter**: a system for automatically filtering inflammatory and outrageous posts within online communities;
2. **Social Media Marketing Tool**: an intelligent web application for managing social media information about products and services through a faceted interface;
3. **Sentic Album**: a content, concept, and context based online personal photo management system;
4. **Sentic Chat**: an IM platform that enriches social communication through semantics and sentics;
5. **Sentic Corner**: an IUI that dynamically collects audio, video, images, and text related to the user's emotions and motions;
6. **Sentic PROMs**: a new framework for measuring health care quality that exploits the ensemble application of standard PROMs and sentic computing.

5.2 Limitations and Future Work

The research carried out in the past few years has laid solid bases for the development of a variety of emotion-sensitive systems and novel applications in the fields of opinion mining and sentiment analysis. One of the main contributions of this book has also been the introduction of a pioneering approach to the analysis of opinions and sentiments, which goes beyond merely keyword-based methods by using common-sense reasoning and linguistics rules. The developed techniques, however, are still far from perfect as the common-sense knowledge base and the list of sentic patterns need to be further extended and the reasoning tools built on top of them, adjusted accordingly. This last section discusses the limitations of such techniques (Sect. 5.2.1) and their further development (Sect. 5.2.2).

5.2.1 Limitations

The validity of the proposed approach mainly depends on the richness of SenticNet. Without a comprehensive resource that encompasses human knowledge, in fact, it is not easy for an opinion-mining system to get a hold of the ability to grasp the cognitive and affective information associated with natural language text and, hence, accordingly aggregate opinions in order to make statistics on them. Attempts to encode human common knowledge are countless and comprehend both resources generated by human experts (or community efforts) and automatically-built knowledge bases. The former kinds of resources are generally too limited, as they need to be handcrafted, the latter too noisy, as they mainly rely on information available on the Web.

The span and the accuracy of knowledge available, however, is not the only limitation of opinion-mining systems. Even though a machine "knows 50 million such things",[1] it needs to be able to accordingly exploit such knowledge through different types of associations, e.g., inferential, causal, analogical, deductive, or inductive. For the purposes of this work, singular value decomposition (SVD) appeared to be a good method for generalizing the information contained in the common-sense knowledge bases, but it is very expensive in both computing time and storage, as it requires costly arithmetic operations such as division and square root in the computation of rotation parameters. This is a big issue as AffectNet is continuing to grow, in parallel with the continuously extended versions of ConceptNet, WNA, and the crowdsourced knowledge coming from GECKA. Moreover, the eigenmoods of AffectiveSpace cannot be easily understood because they are linear combinations of all of the original concept features. Different strategies that clearly show various steps of reasoning might be preferable in the future.

Another limitation of the sentic computing approach is in its typicality. The clearly defined knowledge representation of AffectNet, in fact, does not allow for grasping different concept nuances as the inference of semantic and affective features associated with concepts is bounded. New features associated with a concept can indeed be inferred through the AffectiveSpace process, but the number of new features that can be discovered after reconstructing the concept-feature matrix is limited to the set of features associated with semantically related concepts (that is, concepts that share similar features). However, depending on the context, concepts might need to be associated with features that are not strictly pertinent to germane concepts.

The concept `book`, for example, is typically associated with concepts such as `newspaper` or `magazine`, as it contains knowledge, has pages, etc. In a different context, however, a book could be used as paperweight, doorstop, or even as a weapon. Biased (context-dependent) association of concepts is possible through spectral association, in which spreading activation is concurrently determined by different nodes in the graph representation of AffectNet.

---

[1]http://mitworld.mit.edu/video/484

As concepts considered here are atomic and mono-faceted, it is not easy for the system to grasp the many different ways a concept can be meaningful in a particular context, as the features associated with each concept identify just its typical qualities, traits, or characteristics. Finally, another limitation of the proposed approach is in the lack of time representation. Such an issue is not addressed by any of the currently available knowledge bases, including ConceptNet, upon which AffectNet is built. In the context of sentic computing, however, time representation is not specifically needed as the main aim of the opinion-mining engine is the passage from unstructured natural language data to structured machine-processable information, rather than genuine natural language understanding. Every bag of concepts, in fact, is treated as independent from others in the text data, as the goal is to simply infer a topic and polarity associated with it, rather than understand the whole meaning of the sentence in correlation with adjacent ones. In some cases, however, time representation might be needed for tasks such as comparative opinion analysis and co-reference resolution.

5.2.2 Future Work

In order to overcome some of the above-mentioned limitations, current research work is focusing on expanding AffectNet with different kinds of knowledge (e.g., common-sense, affective knowledge, common knowledge) coming from external resources. Such features are not only useful for improving the accuracy of the opinion-mining engine, but also for reducing the sparseness of the matrix representations of such knowledge bases and, hence, aiding dimensionality reduction procedures.

The set of sentic patterns will continue to be expanded in the future, and new graph-mining and dimensionality-reduction techniques explored to perform reasoning on the common-sense knowledge base. In particular, AffectiveSpace will be built by means of random projections. Further, new classification techniques such as support and relevance vector machines will be experimented, together with the ensemble application of dimensionality reduction and new versions of ELM for emulating fast and affective, cognitive learning and reasoning capabilities and, hence, for jumping to the next NLP curve.

Such a leap, however, is very big: the origins of human language, in fact, has sometimes been termed the hardest problem of science [83]. NLP technologies evolved from the era of punch cards and batch processing (in which the analysis of a natural language sentence could take up to 7 min [245]) to the era of Google and the likes of it (in which millions of webpages can be processed in less than a second). Even the most efficient word-based algorithms, however, perform very poorly, if not properly trained or when contexts and domains change. Such algorithms are limited by the fact that they can process only information they can 'see'. Language, however, is a system where all terms are interdependent and where the value of one is the result of the simultaneous presence of the others [104].

As human text processors, we 'see more than what we see' [103] in which every word activates a cascade of semantically-related concepts that enable the completion of complex NLP tasks, such as word-sense disambiguation, textual entailment, and semantic role labeling, in a quick, seamless and effortless way. Concepts are the glue that holds our mental world together [221]. Without concepts, there would be no mental world in the first place [31]. Needless to say, the ability to organize knowledge into concepts is one of the defining characteristics of the human mind. A truly intelligent system needs physical knowledge of how objects behave, social knowledge of how people interact, sensory knowledge of how things look and taste, psychological knowledge about the way people think, and so on. Having a database of millions of common-sense facts, however, is not enough for computational natural language understanding: we will not only need to teach NLP systems how to handle this knowledge (IQ), but also interpret emotions (EQ) and cultural nuances (CQ).

References

1. Abbasi, A., Chen, H., Salem, A.: Sentiment analysis in multiple languages: feature selection for opinion classification in web forums. ACM Trans. Inf. Syst. **26**(3), 1–34 (2008)
2. Achlioptas, D.: Database-friendly random projections: Johnson-lindenstrauss with binary coins. J. Comput. Syst. Sci. **66**(4), 671–687 (2003)
3. Addis, M., Boch, L., Allasia, W., Gallo, F., Bailer, W., Wright, R.: 100 million hours of audiovisual content: digital preservation and access in the PrestoPRIME project. In: Digital Preservation Interoperability Framework Symposium, Dresden (2010)
4. Agrawal, R., Srikant, R.: Fast algorithm for mining association rules. In: VLDB, Santiago de Chile (1994)
5. von Ahn, L.: Games with a purpose. IEEE Comput. Mag. **6**, 92–94 (2006)
6. von Ahn, L., Dabbish, L.: Labeling images with a computer game. In: CHI, Vienna, pp. 319–326 (2004)
7. von Ahn, L., Ginosar, S., Kedia, M., Liu, R., Blum, M.: Improving accessibility of the web with a computer game. In: CHI, Quebec, pp. 79–82 (2006)
8. von Ahn, L., Kedia, M., Blum, M.: Verbosity: a game for collecting common sense facts. In: CHI, Quebec, pp. 75–78 (2006)
9. von Ahn, L., Liu, R., Blum, M.: Peekaboom: a game for locating objects in images. In: CHI, Quebec, pp. 55–64 (2006)
10. Ailon, N., Chazelle, B.: Faster dimension reduction. Commun. ACM **53**(2), 97–104 (2010)
11. Allen, J.: Natural Language Understanding. Benjamin/Cummings, Menlo Park (1987)
12. Alm, C.O., Roth, D., Sproat, R.: Emotions from text: machine learning for text-based emotion prediction. In: Proceedings of the Conference on Human Language Technology and Empirical Methods in Natural Language Processing, Vancouver, pp. 579–586. Association for Computational Linguistics (2005)
13. Anscombre, J., Ducrot, O.: Deux mais en français. Lingua **43**, 23–40 (1977)
14. Araújo, M., Gonçalves, P., Cha, M., Benevenuto, F.: iFeel: a system that compares and combines sentiment analysis methods. In: Proceedings of the Companion Publication of the 23rd International Conference on World Wide Web Companion, WWW Companion'14, pp. 75–78. International World Wide Web Conferences Steering Committee, Republic and Canton of Geneva, Switzerland (2014)
15. Asher, N., Lascarides, A.: Logics of Conversation. Cambridge University Press, Cambridge (2003)
16. Atassi, H., Esposito, A.: A speaker independent approach to the classification of emotional vocal expressions. In: ICTAI, pp. 147–15 (2008)

© Springer International Publishing Switzerland 2015
E. Cambria, A. Hussain, *Sentic Computing*, Socio-Affective Computing 1,
DOI 10.1007/978-3-319-23654-4

17. Averill, J.R.: A constructivist view of emotion. In: Plutchik, R., Kellerman, H. (eds.) Emotion: Theory, Research and Experience, pp. 305–339. Academic, New York (1980). http://emotion-research.net/biblio/Averill1980
18. Bach, J., Fuller, C., Gupta, A., Hampapur, A., Horowitz, B., Humphrey, R., Jain, R., Shu, C.: Virage image search engine: an open framework for image management. In: Sethi, I., Jain, R. (eds.) Storage and Retrieval for Still Image and Video Databases, vol. 2670, pp. 76–87. SPIE, Bellingham (1996)
19. Baker, C., Fillmore, C., Lowe, J.: The Berkeley FrameNet project. In: COLING/ACL, Montreal, pp. 86–90 (1998)
20. Balahur, A., Hermida, J.M., Montoyo, A.: Building and exploiting emotinet, a knowledge base for emotion detection based on the appraisal theory model. IEEE Trans. Affect. Comput. **3**(1), 88–101 (2012)
21. Balduzzi, D.: Randomized co-training: from cortical neurons to machine learning and back again. arXiv preprint arXiv:1310.6536 (2013)
22. Barrett, L.: Solving the emotion paradox: categorization and the experience of emotion. Personal. Soc. Psychol. Rev. **10**(1), 20–46 (2006)
23. Barrington, L., O'Malley, D., Turnbull, D., Lanckriet, G.: User-centered design of a social game to tag music. In: ACM SIGKDD, Paris, pp. 7–10 (2009)
24. Barwise, J.: An introduction to first-order logic. In: Barwise, J. (ed.) Handbook of Mathematical Logic. Studies in Logic and the Foundations of Mathematics. North-Holland, Amsterdam (1977, 1982). ISBN 978-0-444-86388-1.
25. Beaver, D.: Presupposition and Assertion in Dynamic Semantics. CSLI Publications, Stanford (2008)
26. Benson, T., Sizmur, S., Whatling, J., Arikan, S., McDonald, D., Ingram, D.: Evaluation of a new short generic measure of health status. Inform. Prim. Care **18**(2), 89–101 (2010)
27. Bergner, M., Bobbitt, R., Kressel, S., Pollard, W., Gilson, B., Morris, J.: The sickness impact profile: conceptual formulation and methodology for the development of a health status measure. Int. J. Health Serv. **6**, 393–415 (1976)
28. Bianchi-Berthouze, N.: K-DIME: an affective image filtering system. IEEE Multimed. **10**(3), 103–106 (2003)
29. Bingham, E., Mannila, H.: Random projection in dimensionality reduction: applications to image and text data. In: Proceedings of the Seventh ACM SIGKDD International Conference on Knowledge Discovery and Data Mining, San Francisco, pp. 245–250. ACM (2001)
30. Blitzer, J., Dredze, M., Pereira, F.: Biographies, bollywood, boom-boxes and blenders: domain adaptation for sentiment classification. In: ACL, Prague, vol. 7, pp. 440–447 (2007)
31. Bloom, P.: Glue for the mental world. Nature **421**, 212–213 (2003)
32. Bonanno, G., Papa, A., O'Neill, K., Westphal, M., Coifman, K.: The importance of being flexible: the ability to enhance and suppress emotional expressions predicts long-term adjustment. Psychol. Sci. **15**, 482–487 (2004)
33. Bradford Cannon, W.: Bodily Changes in Pain, Hunger, Fear and Rage: An Account of Recent Researches into the Function of Emotional Excitement. Appleton Century Crofts, New York/London (1915)
34. Bravo-Marquez, F., Mendoza, M., Poblete, B.: Meta-level sentiment models for big social data analysis. Knowl.-Based Syst. **69**, 86–99 (2014)
35. Broca, P.: Anatomie comparée des circonvolutions cérébrales: Le grand lobe limbique. Rev. Anthropol **1**, 385–498 (1878)
36. Brooks, R.: EuroQoL – the current state of play. Health Policy **37**, 53–72 (1996)
37. Burke, A., Heuer, F., Reisberg, D.: Remembering emotional events. Mem. Cognit. **20**, 277–290 (1992)
38. Burkhardt, F., Paeschke, A., Rolfes, M., Sendlmeier, W., Weiss, B.: A database of German emotional speech. In: Interspeech, Lisboa, pp. 1517–1520 (2005)
39. Cahill, L., McGaugh, J.: A novel demonstration of enhanced memory associated with emotional arousal. Conscious. Cognit. **4**(4), 410–421 (1995)
40. Calvo, M., Nummenmaa, L.: Processing of unattended emotional visual scenes. J. Exp. Psychol. Gen. **136**, 347–369 (2007)

41. Calvo, R., D'Mello, S.: Affect detection: an interdisciplinary review of models, methods, and their applications. IEEE Trans. Affect. Comput. **1**(1), 18–37 (2010)
42. Cambria, E., Benson, T., Eckl, C., Hussain, A.: Sentic PROMs: application of sentic computing to the development of a novel unified framework for measuring health-care quality. Expert Syst. Appl. **39**(12), 10533–10543 (2012)
43. Cambria, E., Chandra, P., Sharma, A., Hussain, A.: Do not feel the trolls. In: ISWC, Shanghai (2010)
44. Cambria, E., Fu, J., Bisio, F., Poria, S.: AffectiveSpace 2: enabling affective intuition for concept-level sentiment analysis. In: AAAI, Austin, pp. 508–514 (2015)
45. Cambria, E., Gastaldo, P., Bisio, F., Zunino, R.: An ELM-based model for affective analogical reasoning. Neurocomputing **149**, 443–455 (2015)
46. Cambria, E., Grassi, M., Hussain, A., Havasi, C.: Sentic computing for social media marketing. Multimed. Tools Appl. **59**(2), 557–577 (2012)
47. Cambria, E., Howard, N., Hsu, J., Hussain, A.: Sentic blending: scalable multimodal fusion for continuous interpretation of semantics and sentics. In: IEEE SSCI, Singapore, pp. 108–117 (2013)
48. Cambria, E., Huang, G.B., et al.: Extreme learning machines. IEEE Intell. Syst. **28**(6), 30–59 (2013)
49. Cambria, E., Hussain, A.: Sentic album: content-, concept-, and context-based online personal photo management system. Cognit. Comput. **4**(4), 477–496 (2012)
50. Cambria, E., Hussain, A.: Sentic Computing: Techniques, Tools, and Applications. Springer, Dordrecht (2012)
51. Cambria, E., Hussain, A., Durrani, T., Havasi, C., Eckl, C., Munro, J.: Sentic computing for patient centered application. In: IEEE ICSP, Beijing, pp. 1279–1282 (2010)
52. Cambria, E., Hussain, A., Durrani, T., Zhang, J.: Towards a chinese common and common sense knowledge base for sentiment analysis. In: Jiang, H., Ding, W., Ali, M., Wu, X. (eds.) Advanced Research in Applied Artificial Intelligence. Lecture Notes in Computer Science, vol. 7345, pp. 437–446. Springer, Berlin/Heidelberg (2012)
53. Cambria, E., Hussain, A., Eckl, C.: Taking refuge in your personal sentic corner. In: IJCNLP, Chiang Mai, pp. 35–43 (2011)
54. Cambria, E., Hussain, A., Havasi, C., Eckl, C.: Common sense computing: from the society of mind to digital intuition and beyond. In: Fierrez, J., Ortega, J., Esposito, A., Drygajlo, A., Faundez-Zanuy, M. (eds.) Biometric ID Management and Multimodal Communication. Lecture Notes in Computer Science, vol. 5707, pp. 252–259. Springer, Berlin/Heidelberg (2009)
55. Cambria, E., Hussain, A., Havasi, C., Eckl, C.: SenticSpace: visualizing opinions and sentiments in a multi-dimensional vector space. In: Setchi, R., Jordanov, I., Howlett, R., Jain, L. (eds.) Knowledge-Based and Intelligent Information and Engineering Systems. Lecture Notes in Artificial Intelligence, vol. 6279, pp. 385–393. Springer, Berlin (2010)
56. Cambria, E., Hussain, A., Havasi, C., Eckl, C., Munro, J.: Towards crowd validation of the UK national health service. In: WebSci, Raleigh (2010)
57. Cambria, E., Livingstone, A., Hussain, A.: The hourglass of emotions. In: Esposito, A., Vinciarelli, A., Hoffmann, R., Muller, V. (eds.) Cognitive Behavioral Systems. Lecture Notes in Computer Science, vol. 7403, pp. 144–157. Springer, Berlin/Heidelberg (2012)
58. Cambria, E., Mazzocco, T., Hussain, A., Eckl, C.: Sentic medoids: organizing affective common sense knowledge in a multi-dimensional vector space. In: Liu, D., Zhang, H., Polycarpou, M., Alippi, C., He, H. (eds.) Advances in Neural Networks. Lecture Notes in Computer Science, vol. 6677, pp. 601–610. Springer, Berlin (2011)
59. Cambria, E., Olsher, D., Kwok, K.: Sentic activation: a two-level affective common sense reasoning framework. In: AAAI, Toronto, pp. 186–192 (2012)
60. Cambria, E., Olsher, D., Kwok, K.: Sentic panalogy: swapping affective common sense reasoning strategies and foci. In: CogSci, Sapporo, pp. 174–179 (2012)
61. Cambria, E., Olsher, D., Rajagopal, D.: SenticNet 3: a common and common-sense knowledge base for cognition-driven sentiment analysis. In: AAAI, pp. 1515–1521. Quebec City, (2014)

62. Cambria, E., Rajagopal, D., Kwok, K., Sepulveda, J.: GECKA: game engine for commonsense knowledge acquisition. In: FLAIRS, Hollywood, pp. 282–287 (2015)
63. Cambria, E., Schuller, B., Liu, B., Wang, H., Havasi, C.: Knowledge-based approaches to concept-level sentiment analysis. IEEE Intell. Syst. **28**(2), 12–14 (2013)
64. Cambria, E., Schuller, B., Liu, B., Wang, H., Havasi, C.: Statistical approaches to concept-level sentiment analysis. IEEE Intell. Syst. **28**(3), 6–9 (2013)
65. Cambria, E., Schuller, B., Xia, Y.: New avenues in opinion mining and sentiment analysis. IEEE Intell. Syst. **28**(2), 15–21 (2013)
66. Cambria, E., Wang, H., White, B.: Guest editorial: big social data analysis. Knowl.-Based Syst. **69**, 1–2 (2014)
67. Cambria, E., White, B.: Jumping NLP curves: a review of natural language processing research. IEEE Comput. Intell. Mag. **9**(2), 48–57 (2014)
68. Cambria, E., Xia, Y., Hussain, A.: Affective common sense knowledge acquisition for sentiment analysis. In: LREC, Istanbul, pp. 3580–3585 (2012)
69. Caridakis, G., Castellano, G., Kessous, L., Raouzaiou, A., Malatesta, L., Asteriadis, S., Karpouzis, K.: Multimodal emotion recognition from expressive faces, body gestures and speech. In: Artificial intelligence and innovations 2007: from theory to applications, Athens, pp. 375–388 (2007)
70. Castellano, G., Kessous, L., Caridakis, G.: Multimodal emotion recognition from expressive faces, body gestures and speech. In: Doctoral Consortium of ACII, Lisbon (2007)
71. Chaiken, S., Trope, Y.: Dual-Process Theories in Social Psychology. Guilford, New York (1999)
72. Chandra, P., Cambria, E., Hussain, A.: Clustering social networks using interaction semantics and sentics. In: Wang, J., Yen, G., Polycarpou, M. (eds.) Advances in Neural Networks. Lecture Notes in Computer Science, vol. 7367, pp. 379–385. Springer, Heidelberg (2012)
73. Chandra, P., Cambria, E., Pradeep, A.: Enriching social communication through semantics and sentics. In: IJCNLP, Chiang Mai, pp. 68–72 (2011)
74. Chang, H.: Emotion barometer of reading: user interface design of a social cataloging website. In: International Conference on Human Factors in Computing Systems, Boston (2009)
75. Chaumartin, F.R.: Upar7: a knowledge-based system for headline sentiment tagging. In: Proceedings of the 4th International Workshop on Semantic Evaluations, Prague, pp. 422–425. Association for Computational Linguistics (2007)
76. Chen, L.S.H.: Joint processing of audio-visual information for the recognition of emotional expressions in human-computer interaction. Ph.D. thesis, Citeseer (2000)
77. Chenlo, J.M., Losada, D.E.: An empirical study of sentence features for subjectivity and polarity classification. Inf. Sci. **280**, 275–288 (2014)
78. Chi, P., Lieberman, H.: Intelligent assistance for conversational storytelling using story patterns. In: IUI, Palo Alto (2011)
79. Chikersal, P., Poria, S., Cambria, E.: SeNTU: sentiment analysis of tweets by combining a rule-based classifier with supervised learning. In: Proceedings of the International Workshop on Semantic Evaluation (SemEval-2015), Denver (2015)
80. Chikersal, P., Poria, S., Cambria, E., Gelbukh, A., Siong, C.E.: Modelling public sentiment in twitter: using linguistic patterns to enhance supervised learning. In: Computational Linguistics and Intelligent Text Processing, pp. 49–65. Springer (2015)
81. Chklovski, T.: Learner: a system for acquiring commonsense knowledge by analogy. In: K-CAP, Sanibel Island, pp. 4–12 (2003)
82. Chomsky, N.: Three models for the description of language. IRE Trans. Inf. Theory **2**(3), 113–124 (1956)
83. Christiansen, M., Kirby, S.: Language evolution: the hardest problem in science? In: Christiansen, M., Kirby, S. (eds.) Language Evolution, chap. 1, pp. 1–15. Oxford University Press, Oxford (2003)
84. Christianson, S., Loftus, E.: Remembering emotional events: the fate of detailed information. Cognit. Emot. **5**, 81–108 (1991)

85. Chung, J.K.C., Wu, C.E., Tsai, R.T.H.: Improve polarity detection of online reviews with bag-of-sentimental-concepts. In: Proceedings of the 11th ESWC. Semantic Web Evaluation Challenge, Crete. Springer (2014)
86. Cochrane, T.: Eight dimensions for the emotions. Soc. Sci. Inf. **48**(3), 379–420 (2009)
87. Codd, E.: A relational model of data for large shared data banks. Commun. ACM **13**(6), 377–387 (1970)
88. Codd, E.: Further normalization of the data base relational model. Tech. rep., IBM Research Report, New York (1971)
89. Codd, E.: Recent investigations into relational data base systems. Tech. Rep. RJ1385, IBM Research Report, New York (1974)
90. Coppock, E., Beaver, D.: Principles of the exclusive muddle. J. Semant. (2013). doi:10.1093/jos/fft007
91. Cowie, R., Douglas-Cowie, E.: Automatic statistical analysis of the signal and prosodic signs of emotion in speech. In: Proceedings of the Fourth International Conference on Spoken Language (ICSLP 96), Philadelphia, vol. 3, pp. 1989–1992. IEEE (1996)
92. Csikszentmihalyi, M.: Flow: The Psychology of Optimal Experience. Harper Perennial, San Francisco (1991)
93. Culyer, A., Lavers, R., Williams, A.: Social indicators: Health. Soc. Trends **2**, 31–42 (1971)
94. Dalgleish, T.: The emotional brain. Nat. Perspect. **5**, 582–589 (2004)
95. Dalgleish, T., Dunn, B., Mobbs, D.: Affective neuroscience: past, present, and future. Emotion Review **1.4** (2009): 355–368. (2009)
96. Damasio, A.: Descartes' Error: Emotion, Reason, and the Human Brain. Grossett/Putnam, New York (1994)
97. Damasio, A.: Looking for Spinoza: Joy, Sorrow, and the Feeling Brain. Harcourt, Inc., Orlando (2003)
98. Darwin, C.: The Expression of the Emotions in Man and Animals. John Murray, London (1872)
99. Datcu, D., Rothkrantz, L.: Semantic audio-visual data fusion for automatic emotion recognition. In: Euromedia, Citeseer (2008)
100. Date, C., Darwen, H.: A Guide to the SQL Standard. Addison-Wesley, Reading (1993)
101. Datta, R., Wang, J.: ACQUINE: aesthetic quality inference engine – real-time automatic rating of photo aesthetics. In: International Conference on Multimedia Information Retrieval, Philadelphia (2010)
102. Davidov, D., Tsur, O., Rappoport, A.: Enhanced sentiment learning using twitter hashtags and smileys. In: Proceedings of the 23rd International Conference on Computational Linguistics: Posters, Iaşi, pp. 241–249. Association for Computational Linguistics (2010)
103. Davidson, D.: Seeing through language. R. Inst. Philos. Suppl. **42**, 15–28 (1997)
104. De Saussure, F.: Cours de linguistique générale. Payot, Paris (1916)
105. Decherchi, S., Gastaldo, P., Zunino, R., Cambria, E., Redi, J.: Circular-ELM for the reduced-reference assessment of perceived image quality. Neurocomputing **102**, 78–89 (2013)
106. Dellaert, F., Polzin, T., Waibel, A.: Recognizing emotion in speech. In: Proceedings of the Fourth International Conference on Spoken Language (ICSLP 96), Philadelphia, vol. 3, pp. 1970–1973. IEEE (1996)
107. Di Fabbrizio, G., Aker, A., Gaizauskas, R.: Starlet: multi-document summarization of service and product reviews with balanced rating distributions. In: ICDM SENTIRE, Vancouver, pp. 67–74 (2011)
108. Donabedian, A.: Evaluating the quality of medical care. The Millbank Meml. Fund Quart. **44**, 166–203 (1966)
109. Douglas-Cowie, E.: Humaine deliverable D5g: mid term report on database exemplar progress. Tech. rep., Information Society Technologies (2006)
110. Dragoni, M., Tettamanzi, A.G., da Costa Pereira, C.: A fuzzy system for concept-level sentiment analysis. In: Semantic Web Evaluation Challenge, pp. 21–27. Springer, Cham (2014)

111. Duthil, B., Trousset, F., Dray, G., Montmain, J., Poncelet, P.: Opinion extraction applied to criteria. In: Database and Expert Systems Applications, pp. 489–496. Springer, Heidelberg/New York (2012)
112. Dyer, M.: Connectionist natural language processing: a status report. In: Computational architectures integrating neural and symbolic processes, vol. 292, pp. 389–429. Kluwer Academic, Dordrecht (1995)
113. Eckart, C., Young, G.: The approximation of one matrix by another of lower rank. Psychometrika **1**(3), 211–218 (1936)
114. Ekman, P.: Universal facial expressions of emotion. In: Culture and Personality: Contemporary Readings. Aldine, Chicago (1974)
115. Ekman, P., Dalgleish, T., Power, M.: Handbook of Cognition and Emotion. Wiley, Chichester (1999)
116. Ekman, P., Friesen, W.: Facial Action Coding System: A Technique for the Measurement of Facial Movement. Consulting Psychologists Press, Palo Alto (1978)
117. Elliott, C.D.: The affective reasoner: a process model of emotions in a multi-agent system. Ph.D. thesis, Northwestern University, Evanston (1992)
118. Ephron, H.: 1001 Books for Every Mood: A Bibliophile's Guide to Unwinding, Misbehaving, Forgiving, Celebrating, Commiserating. Adams Media, Avon (2008)
119. Epstein, S.: Cognitive-experiential self-theory of personality. In: Millon, T., Lerner, M. (eds.) Comprehensive Handbook of Psychology, vol. 5, pp. 159–184. Wiley, Hoboken (2003)
120. Ernest, D.: Representations of Commonsense Knowledge. Morgan Kaufmann, San Mateo (1990)
121. Esuli, A., Sebastiani, F.: Sentiwordnet: a publicly available lexical resource for opinion mining. In: Proceedings of LREC, Genoa, vol. 6, pp. 417–422 (2006)
122. Eyben, F., Wollmer, M., Schuller, B.: OpenEAR—introducing the munich open-source emotion and affect recognition toolkit. In: 3rd International Conference on Affective Computing and Intelligent Interaction and Workshops (ACII 2009), Amsterdam, pp. 1–6. IEEE (2009)
123. Fanshel, S., Bush, J.: A health status index and its application to health-services outcomes. Oper. Res. **18**, 1021–1066 (1970)
124. Fauconnier, G., Turner, M.: The Way We Think: Conceptual Blending and the Mind's Hidden Complexities. Basic Books, New York (2003)
125. Fellbaum, C.: WordNet: An Electronic Lexical Database (Language, Speech, and Communication). The MIT Press, Cambridge (1998)
126. Flickner, M., Sawhney, H., Niblack, W., Ashley, J., Huang, Q., Dom, B., Gorkani, M., Hafner, J., Lee, D., Petkovic, D., Steele, D., Yanker, P.: Query by image and video content: the QBIC system. Computer **28**(9), 23–32 (1995)
127. Fontaine, J., Scherer, K., Roesch, E., Ellsworth, P.: The world of emotions is not two-dimensional. Psychol. Sci. **18**(12), 1050–1057 (2007)
128. Frankel, C., Swain, M.J., Athitsos, V.: WebSeer: an image search engine for the world wide web. Tech. rep., University of Chicago (1996)
129. Freitas, A., Castro, E.: Facial expression: the effect of the smile in the treatment of depression. empirical study with Portuguese subjects. In: Freitas-Magalhães, A. (ed.) Emotional Expression: The Brain and The Face, pp. 127–140. University Fernando Pessoa Press, Porto (2009)
130. Friesen, W.V., Ekman, P.: Emfacs-7: Emotional facial action coding system. Unpublished manuscript, University of California at San Francisco, vol. 2 (1983)
131. Frijda, N.: The laws of emotions. Am. Psychol. **43**(5) (1988)
132. Gezici, G., Dehkharghani, R., Yanikoglu, B., Tapucu, D., Saygin, Y.: Su-sentilab: a classification system for sentiment analysis in Twitter. In: Proceedings of the International Workshop on Semantic Evaluation, Atlanta, pp. 471–477 (2013)
133. Glorot, X., Bordes, A., Bengio, Y.: Domain adaptation for large-scale sentiment classification: a deep learning approach. In: ICML, Bellevue (2011)
134. Goertzel, B., Silverman, K., Hartley, C., Bugaj, S., Ross, M.: The Baby Webmind project. In: AISB, Birmingham (2000)

135. Grassi, M., Cambria, E., Hussain, A., Piazza, F.: Sentic web: a new paradigm for managing social media affective information. Cognit. Comput. **3**(3), 480–489 (2011)
136. Gunes, H., Piccardi, M.: Bi-modal emotion recognition from expressive face and body gestures. J. Netw. Comput. Appl. **30**(4), 1334–1345 (2007)
137. Gupta, R., Kochenderfer, M., Mcguinness, D., Ferguson, G.: Common sense data acquisition for indoor mobile robots. In: AAAI, San Jose, pp. 605–610 (2004)
138. Hacker, S., von Ahn, L.: Matchin: eliciting user preferences with an online game. In: CHI, Boston, pp. 1207–1216 (2009)
139. Hanjalic, A.: Extracting moods from pictures and sounds: towards truly personalized TV. IEEE Signal Process. Mag. **23**(2), 90–100 (2006)
140. Hatzivassiloglou, V., McKeown, K.: Predicting the semantic orientation of adjectives. In: ACL/EACL, Madrid (1997)
141. Havasi, C.: Discovering semantic relations using singular value decomposition based techniques. Ph.D. thesis, Brandeis University (2009)
142. Havasi, C., Speer, R., Alonso, J.: ConceptNet 3: a flexible, multilingual semantic network for common sense knowledge. In: RANLP, Borovets (2007)
143. Havasi, C., Speer, R., Holmgren, J.: Automated color selection using semantic knowledge. In: AAAI CSK, Arlington (2010)
144. Herdagdelen, A., Baroni, M.: The concept game: better commonsene knowledge extraction by combining text mining and game with a purpose. In: AAAI CSK, Arlington (2010)
145. Heyting, A.: Intuitionism. An introduction. North-Holland, Amsterdam (1956)
146. Horsman, J., Furlong, W., Feeny, D., Torrance, G.: The health utility index (HUI): concepts, measurement, properties and applications. Health Qual. Life Outcomes **1**(54), 1–13 (2003)
147. Howard, N., Cambria, E.: Intention awareness: improving upon situation awareness in human-centric environments. Hum.-Centric Comput. Inf. Sci. **3**(9), 1–17 (2013)
148. Hu, M., Liu, B.: Mining and summarizing customer reviews. In: KDD, Seattle (2004)
149. Hu, M., Liu, B.: Mining and summarizing customer reviews. In: Proceedings of the tenth ACM SIGKDD International Conference on Knowledge Discovery and Data Mining, Seattle, pp. 168–177. ACM (2004)
150. Huang, G.B.: An insight into extreme learning machines: random neurons, random features and kernels. Cognit. Comput. **6**(3), 376–390 (2014)
151. Huang, G.B., Cambria, E., Toh, K.A., Widrow, B., Xu, Z.: New trends of learning in computational intelligence. IEEE Comput. Intell. Mag. **10**(2), 16–17 (2015)
152. Huang, G.B., Chen, L., Siew, C.K.: Universal approximation using incremental constructive feedforward networks with random hidden nodes. IEEE Trans. Neural Netw. **17**(4), 879–892 (2006)
153. Huang, G.B., Wang, D.H., Lan, Y.: Extreme learning machines: a survey. Int. J. Mach. Learn. Cybern. **2**(2), 107–122 (2011)
154. Huang, G.B., Zhou, H., Ding, X., Zhang, R.: Extreme learning machine for regression and multiclass classification. IEEE Trans. Syst. Man Cybern. Part B: Cybern. **42**(2), 513–529 (2012)
155. Huang, J., Ravi, S., Mitra, M., Zhu, W., Zabih, R.: Image indexing using color correlograms. In: IEEE CVPR, San Juan, pp. 762–768 (1997)
156. Imparato, N., Harari, O.: Jumping the Curve: Innovation and Strategic Choice in an Age of Transition. Jossey-Bass Publishers, San Francisco (1996)
157. James, W.: What is an emotion? Mind **34**, 188–205 (1884)
158. Jayez, J., Winterstein, G.: Additivity and probability. Lingua **132**(85–102) (2013)
159. Jing, F., Wang, C., Yao, Y., Deng, K., Zhang, L., Ma, W.Y.: IGroup: web image search results clustering. In: ACM Multimedia, Santa Barbara (2006)
Johnstone, T.: Emotional speech elicited using computer games. In: Proceedings of the Fourth International Conference on Spoken Language (ICSLP 96), Philadelphia, vol. 3, pp. 1985-1988. IEEE (1996)
160. Johnstone, T.: Emotional speech elicited using computer games. In: Proceedings of the Fourth International Conference on Spoken Language (ICSLP 96), Philadelphia, vol. 3, pp. 1985–1988. IEEE (1996)

161. Joshi, M., Rose, C.: Generalizing dependency features for opinion mining. In: ACL/IJCNLP, Singapore (2009)
162. Kalchbrenner, N., Grefenstette, E., Blunsom, P.: A convolutional neural network for modelling sentences. CoRR abs/1404.2188 (2014)
163. Kamps, J., Marx, M., Mokken, R., de Rijke, M.: Using WordNet to measure semantic orientation of adjectives. In: LREC, Lisbon, pp. 1115–1118 (2004)
164. Kapoor, A., Burleson, W., Picard, R.: Automatic prediction of frustration. Int. J. Hum.-Comput. Stud. **65**, 724–736 (2007)
165. Karttunen, L.: Presuppositions of compound sentences. Linguist. Inq. **4**(2), 169–193 (1973)
166. Keelan, B.: Handbook of Image Quality. Marcel Dekker, New York (2002)
167. Kenji, M.: Recognition of facial expression from optical flow. IEICE Trans. Inf. Syst. **74**(10), 3474–3483 (1991)
168. Kim, S., Hovy, E.: Automatic detection of opinion bearing words and sentences. In: IJCNLP, Jeju Island, pp. 61–66 (2005)
169. Kim, S., Hovy, E.: Extracting opinions, opinion holders, and topics expressed in online news media text. In: Workshop on Sentiment and Subjectivity in Text, Sydney (2006)
170. Kirkpatrick, L., Epstein, S.: Cognitive experiential self-theory and subjective probability: further evidence for two conceptual systems. J. Personal. Soc. Psychol. **63**, 534–544 (1992)
171. Kouloumpis, E., Wilson, T., Moore, J.: Twitter sentiment analysis: the good the bad and the omg! ICWSM **11**, 538–541 (2011)
172. Krumhuber, E., Kappas, A.: Moving smiles: the role of dynamic components for the perception of the genuineness of smiles. J. Nonverbal Behav. **29**(1), 3–24 (2005)
173. Kuo, Y., Lee, J., Chiang, K., Wang, R., Shen, E., Chan, C., Hu, J.Y.: Community-based game design: experiments on social games for commonsense data collection. In: ACM SIGKDD, Paris, pp. 15–22 (2009)
174. Lacy, L.: OWL: Representing Information Using the Web Ontology Language. Trafford Publishing, Victoria (2005)
175. Lakoff, G.: Women, Fire, and Dangerous Things. University Of Chicago Press, Chicago (1990)
176. Lanczos, C.: An iteration method for the solution of the eigenvalue problem of linear differential and integral operators. J. Res. Natl. Bur. Stand. **45**(4), 255–282 (1950)
177. Laney, C., Campbell, H., Heuer, F., Reisberg, D.: Memory for thematically arousing events. Mem. Cognit. **32**(7), 1149–1159 (2004)
178. Lanitis, A., Taylor, C.J., Cootes, T.F.: A unified approach to coding and interpreting face images. In: Proceedings or the Fifth International Conference on Computer Vision, Boston, pp. 368–373. IEEE (1995)
179. Lansdale, M., Edmonds, E.: Using memory for events in the design of personal filing systems. Int. J. Man-Mach. Stud. **36**(1), 97–126 (1992)
180. Law, E., von Ahn, L., Dannenberg, R., Crawford, M.: Tagatune: a game for music and sound annotation. In: International Conference on Music Information Retrieval, Vienna, pp. 361–364 (2007)
181. Lazarus, R.: Emotion and Adaptation. Oxford University Press, New York (1991)
182. Ledoux, J.: Synaptic Self. Penguin Books, New York (2003)
183. Berners-Lee, T., Hendler, J., Lassila, O.: The semantic web. Sci. Am. **284**(5), 28–37 (2001)
184. Lee, H., Grosse, R., Ranganath, R., Ng, A.Y.: Unsupervised learning of hierarchical representations with convolutional deep belief networks. Commun. ACM **54**(10), 95–103 (2011)
185. Lempel, R., Soffer, A.: PicASHOW: pictorial authority search by hyperlinks on the web. In: WWW, Hong Kong (2001)
186. Lenat, D., Guha, R.: Building Large Knowledge-Based Systems: Representation and Inference in the Cyc Project. Addison-Wesley, Boston (1989)
187. Lew, M., Sebe, N., Djeraba, C., Jain, R.: Content-based multimedia information retrieval: state of the art and challenges. ACM Trans. Multimed. Comput. Commun. Appl. **2**(1), 1–19 (2006)
188. Lewis, M.: Self-conscious emotions: embarrassment, pride, shame, and guilt. In: Handbook of Cognition and Emotion, vol. 2, pp. 623–636. Guilford Press, Chichester (2000)

189. Lewis, M., Granic, I.: Emotion, Development, and Self-Organization: Dynamic Systems Approaches to Emotional Development. Cambridge University Press, Cambridge (2002)
190. Lieberman, H., Rosenzweig, E., Singh, P.: ARIA: an agent for annotating and retrieving images. IEEE Comput. **34**(7), 57–62 (2001)
191. Lieberman, H., Selker, T.: Out of context: computer systems that adapt to, and learn from, context. IBM Syst. J. **39**(3), 617–632 (2000)
192. Lieberman, M.: Social cognitive neuroscience: a review of core processes. Ann. Rev. Psychol. **58**, 259–89 (2007)
193. Lin, K.H.Y., Yang, C., Chen, H.H.: What emotions do news articles trigger in their readers? In: Proceedings of the 30th Annual International ACM SIGIR Conference on Research and Development in Information Retrieval, pp. 733–734. ACM (2007)
194. Lin, Z., Ng, H.T., Kan, M.Y.: A PDTB-styled end-to-end discourse parser. Nat. Lang. Eng. **20**(2), 151–184 (2014)
195. Liu, H., Singh, P.: ConceptNet-a practical commonsense reasoning tool-kit. BT Technol. J. **22**(4), 211–226 (2004)
196. Lu, W., Zeng, K., Tao, D., Yuan, Y., Gao, X.: No-reference image quality assessment in contourlet domain. Neurocomputing **73**(4–6), 784–794 (2012)
197. Lu, Y., Dhillon, P., Foster, D.P., Ungar, L.: Faster ridge regression via the subsampled randomized hadamard transform. In: Burges, C.J.C., Bottou, L., Welling, M., Ghahramani, Z., Weinberger, K.Q. (eds.) Advances in Neural Information Processing Systems, pp. 369–377. Curran Associates, Inc., New York (2013)
198. Ma, H., Chandrasekar, R., Quirk, C., Gupta, A.: Page hunt: improving search engines using human computation games. In: SIGIR, Boston, pp. 746–747 (2009)
199. Machajdik, J., Hanbury, A.: Affective image classification using features inspired by psychology and art theory. In: International Conference on Multimedia, Florence (2010)
200. Maclean, P.: Psychiatric implications of physiological studies on frontotemporal portion of limbic system. Electroencephalogr Clin Neurophysiol Suppl **4**, 407–18 (1952)
201. Magritte, R.: Les mots et les images. La Révolution surréaliste **12** (1929)
202. Manning, C.: Part-of-speech tagging from 97 % to 100 %: Is it time for some linguistics? In: Gelbukh, A. (ed.) Computational Linguistics and Intelligent Text Processing. Lecture Notes in Computer Science, vol. 6608, pp. 171–189. Springer, Berlin (2011)
203. Mansoorizadeh, M., Charkari, N.M.: Multimodal information fusion application to human emotion recognition from face and speech. Multimed. Tools Appl. **49**(2), 277–297 (2010)
204. Markotschi, T., Volker, J.: GuessWhat?! – Human intelligence for mining linked data. In: EKAW, Lisbon (2010)
205. Matsumoto, D.: More evidence for the universality of a contempt expression. Motiv. Emot. **16**(4), 363–368 (1992)
206. McCarthy, J.: Programs with common sense. In: Teddington Conference on the Mechanization of Thought Processes (1959)
207. McClelland, J.: Is a Machine realization of truly human-like intelligence achievable? Cogn Comput. **1**,17–21 (2009)
208. Mehrabian, A.: Pleasure-arousal-dominance: a general framework for describing and measuring individual differences in temperament. Current Psychol. **14**(4), 261–292 (1996)
209. Melville, P., Gryc, W., Lawrence, R.D.: Sentiment analysis of blogs by combining lexical knowledge with text classification. In: ACM SIGKDD International Conference on Knowledge Discovery and Data Mining, Paris, pp. 1275–1284. ACM (2009)
210. Menon, A.K., Elkan, C.: Fast algorithms for approximating the singular value decomposition. ACM Trans. Knowl. Discov. Data (TKDD) **5**(2), 13 (2011)
211. Milewski, A., Smith, T.: Providing presence cues to telephone users. In: ACM Conference on Computer Supported Cooperative Work (2000)
212. Minsky, M.: The Society of Mind. Simon and Schuster, New York (1986)
213. Minsky, M.: Commonsense-based interfaces. Commun. ACM **43**(8), 67–73 (2000)
214. Minsky, M.: The Emotion Machine: Commonsense Thinking, Artificial Intelligence, and the Future of the Human Mind. Simon & Schuster, New York (2006)

215. Mishne, G.: Experiments with mood classification in blog posts. In: Proceedings of ACM SIGIR 2005 Workshop on Stylistic Analysis of Text for Information Access, vol. 19 (2005)
216. Mohammad, S.M., Kiritchenko, S., Zhu, X.: NRC-Canada: building the state-of-the-art in sentiment analysis of tweets. In: SemEval, Atlanta, pp. 321–327 (2013)
217. Morency, L.P., Mihalcea, R., Doshi, P.: Towards multimodal sentiment analysis: harvesting opinions from the web. In: Proceedings of the 13th international conference on multimodal interfaces, pp. 169–176. ACM, New York, (2011)
218. Morrison, D., Maillet, S., Bruno, E.: Tagcaptcha: annotating images with captchas. In: ACM SIGKDD, Paris, pp. 44–45 (2009)
219. Mueller, E.: Natural Language Processing with ThoughtTreasure. Signifonn, New York (1998)
220. Mueller, E.: Commonsense Reasoning. Morgan Kaufmann (2006)
221. Murphy, G.: The Big Book of Concepts. The MIT Press, Cambridge (2004)
222. Murray, I.R., Arnott, J.L.: Toward the simulation of emotion in synthetic speech: a review of the literature on human vocal emotion. J. Acoust. Soc. Am. **93**(2), 1097–1108 (1993)
223. Nakazato, M., Manola, L., Huang, T.: ImageGrouper: search, annotate and organize images by groups. In: Chang, S., Chen, Z., Lee, S. (eds.) Recent Advances in Visual Information Systems. Lecture Notes in Computer Science, vol. 2314, pp. 93–105. Springer, Berlin (2002)
224. Narwaria, M., Lin, W.: Objective image quality assessment based on support vector regression. IEEE Trans. Neural Netw. **12**(3), 515–519 (2010)
225. Navas, E., Hernáez, I., Luengo, I.: An objective and subjective study of the role of semantics and prosodic features in building corpora for emotional TTS. IEEE Trans. Audio Speech Lang. Process. **14**(4), 1117–1127 (2006)
226. Neisser, U.: Cognitive Psychology. Appleton Century Crofts, New York (1967)
227. Nguyen, L., Wu, P., Chan, W., Peng, W., Zhang, Y.: Predicting collective sentiment dynamics from time-series social media. In: KDD WISDOM, Beijing, vol. 6 (2012)
228. O'Hare, N., Lee, H., Cooray, S., Gurrin, C., Jones, G., Malobabic, J., O'Connor, N., Smeaton, A., Uscilowski, B.: MediAssist: using content-based analysis and context to manage personal photo collections. In: CIVR, Tempe, pp. 529–532 (2006)
229. Ohman, A., Soares, J.: Emotional conditioning to masked stimuli: expectancies for aversive outcomes following nonre- cognized fear-relevant stimuli. J. Exp. Psychol. Gen. **127**(1), 69–82 (1998)
230. Ortony, A., Clore, G., Collins, A.: The Cognitive Structure of Emotions. Cambridge University Press, Cambridge (1988)
231. Osgood, C., May, W., Miron, M.: Cross-Cultural Universals of Affective Meaning. University of Illinois Press, Urbana (1975)
232. Osgood, C., Suci, G., Tannenbaum, P.: The Measurement of Meaning. University of Illinois Press, Urbana (1957)
233. Pak, A., Paroubek, P.: Twitter as a corpus for sentiment analysis and opinion mining. In: LREC, Valletta, pp. 1320–1326 (2010)
234. Pampalk, E., Rauber, A., Merkl, D.: Content-based organization and visualization of music archives. In: ACM International Conference on Multimedia, Juan les Pins (2002)
235. Pang, B., Lee, L.: A sentimental education: sentiment analysis using subjectivity summarization based on minimum cuts. In: ACL, Barcelona, pp. 271–278 (2004)
236. Pang, B., Lee, L.: Seeing stars: exploiting class relationships for sentiment categorization with respect to rating scales. In: ACL, Ann Arbor, pp. 115–124 (2005)
237. Pang, B., Lee, L.: Opinion mining and sentiment analysis. Found. Trends Inf. Retr. **2**, 1–135 (2008)
238. Pang, B., Lee, L., Vaithyanathan, S.: Thumbs up?: sentiment classification using machine learning techniques. In: EMNLP, Philadelphia, vol. 10, pp. 79–86. ACL (2002)
239. Pang, B., Lee, L., Vaithyanathan, S.: Thumbs up? Sentiment classification using machine learning techniques. In: EMNLP, Philadelphia, pp. 79–86 (2002)
240. Pantic, M.: Affective computing. In: Encyclopedia of Multimedia Technology and Networking, vol. 1, pp. 8–14. Idea Group Reference (2005)

241. Papez, J.: A proposed mechanism of emotion. Neuropsychiatry Clin. Neurosci. **7**, 103–112 (1937)
242. Park, H., Jun, C.: A simple and fast algorithm for k-medoids clustering. Expert Syst. Appl. **36**(2), 3336–3341 (2009)
243. Parrott, W.: Emotions in Social Psychology. Psychology Press, Philadelphia (2001)
244. Pearl, J.: Bayesian networks: a model of self-activated memory for evidential reasoning. Tech. Rep. CSD-850017, UCLA Technical Report, Irvine (1985)
245. Plath, W.: Multiple path analysis and automatic translation. Booth pp. 267–315 (1967)
246. Plutchik, R.: The nature of emotions. Am. Sci. **89**(4), 344–350 (2001)
247. Popescu, A., Etzioni, O.: Extracting product features and opinions from reviews. In: HLT/EMNLP, Vancouver (2005)
248. Poria, S., Cambria, E., Gelbukh, A.: Deep convolutional neural network textual features and multiple kernel learning for utterance-level multimodal sentiment analysis. In: EMNLP, Lisbon, pp. 2539–2544 (2015)
249. Poria, S., Cambria, E., Gelbukh, A., Bisio, F., Hussain, A.: Sentiment data flow analysis by means of dynamic linguistic patterns. IEEE Comput. Intell. Mag. **10**(4), 26–36 (2015)
250. Poria, S., Cambria, E., Howard, N., Huang, G.-B., Hussain, A.: Fusing audio, visual and textual clues for sentiment analysis from multimodal content. Neurocomputing (2015). doi: 10.1016/j.neucom.2015.01.095 (2015)
251. Poria, S., Gelbukh, A., Hussain, A., Howard, A., Das, D., Bandyopadhyay, S.: Enhanced SenticNet with affective labels for concept-based opinion mining. IEEE Intell. Syst. **28**(2), 31–38 (2013)
252. Poria, S., Cambria, E., Hussain, A., Huang, G.B.: Towards an intelligent framework for multimodal affective data analysis. Neural Networks **63**, 104–116 (2015)
253. Poria, S., Cambria, E., Winterstein, G., Huang, G.B.: Sentic patterns: dependency-based rules for concept-level sentiment analysis. Knowl.-Based Syst. **69**, 45–63 (2014)
254. Poria, S., Gelbukh, A., Cambria, E., Das, D., Bandyopadhyay, S.: Enriching SenticNet polarity scores through semi-supervised fuzzy clustering. In: IEEE ICDM, Brussels, pp. 709–716 (2012)
255. Poria, S., Gelbukh, A., Cambria, E., Hussain, A., Huang, G.B.: EmoSenticSpace: a novel framework for affective common-sense reasoning. Knowl.-Based Syst. **69**, 108–123 (2014)
256. Porkaew, K., Chakrabarti, K.: Query refinement for multimedia similarity retrieval in MARS. In: ACM International Conference on Multimedia, pp. 235–238. ACM, New York (1999)
257. Potts, C.: The Logic of Conventional Implicatures. Oxford University Press, Oxford (2005)
258. Prinz, J.: Gut Reactions: A Perceptual Theory of Emotion. Oxford University Press, Oxford (2004)
259. Pudil, P., Ferri, F., Novovicova, J., Kittler, J.: Floating search methods for feature selection with nonmonotonic criterion functions. In: IAPR, Jerusalem, pp. 279–283 (1994)
260. Pun, T., Alecu, T.I., Chanel, G., Kronegg, J., Voloshynovskiy, S.: Brain-computer interaction research at the computer vision and multimedia laboratory, university of Geneva. IEEE Trans. Neural Syst. Rehabil. Eng. **14**(2), 210–213 (2006)
261. Qazi, A., Raj, R.G., Tahir, M., Cambria, E., Syed, K.B.S.: Enhancing business intelligence by means of suggestive reviews. Sci. World J. **2014**, 1–11 (2014)
262. Qi, H., Wang, X., Iyengar, S.S., Chakrabarty, K.: Multisensor data fusion in distributed sensor networks using mobile agents. In: Proceedings of 5th International Conference on Information Fusion, Annapolis, pp. 11–16 (2001)
263. Rajagopal, D., Cambria, E., Olsher, D., Kwok, K.: A graph-based approach to commonsense concept extraction and semantic similarity detection. In: WWW, Rio De Janeiro, pp. 565–570 (2013)
264. Rao, D., Ravichandran, D.: Semi-supervised polarity lexicon induction. In: EACL, Athens, pp. 675–682 (2009)
265. Recupero, D.R., Presutti, V., Consoli, S., Gangemi, A., Nuzzolese, A.G.: Sentilo: frame-based sentiment analysis. Cognit. Comput. **7**(2), 211–225 (2014)

266. Redi, J., Gastaldo, P., Heynderickx, I., Zunino, R.: Color distribution information for the reduced-reference assessment of perceived image quality. IEEE Trans. Circuits Syst. Video Technol. **20**(12), 1757–1769 (2012)
267. Reisberg, D., Heuer, F.: Memory for emotional events. In: Reisberg, D., Hertel, P. (eds.) Memory and Emotion, pp. 3–41. Oxford University Press, New York (2004)
268. Reiter, R.: A logic for default reasoning. Artif. Intell. **13**, 81–132 (1980)
269. Repp, S.: Negation in Gapping. Oxford University Press, Oxford (2009)
270. Richards, J., Butler, E., Gross, J.: Emotion regulation in romantic relationships: the cognitive consequences of concealing feelings. J. Soc. Personal Relatsh. **20**, 599–620 (2003)
271. Ridella, S., Rovetta, S., Zunino, R.: Circular backpropagation networks for classification. IEEE Trans. Neural Netw. **8**(1), 84–97 (1997)
272. Riloff, E., Wiebe, J.: Learning extraction patterns for subjective expressions. In: EMNLP, Sapporo, pp. 105–112 (2003)
273. Riloff, E., Wiebe, J.: Learning extraction patterns for subjective expressions. In: Proceedings of the 2003 conference on empirical methods in natural language processing, pp. 105–112. Association for Computational Linguistics (2003)
274. Rowe, M., Butters, J.: Assessing trust: contextual accountability. In: ESWC, Heraklion (2009)
275. Russell, J.: Affective space is bipolar. J. Personal. Soc. Psychol. **37**, 345–356 (1979)
276. Russell, J.: Core affect and the psychological construction of emotion. Psychol. Rev. **110**, 145–172 (2003)
277. dos Santos, C.N., Gatti, M.: Deep convolutional neural networks for sentiment analysis of short texts. In: Proceedings of the 25th International Conference on Computational Linguistics (COLING), Dublin (2014)
278. Saragih, J.M., Lucey, S., Cohn, J.F.: Face alignment through subspace constrained mean-shifts. In: IEEE 12th International Conference on Computer Vision, Kyoto, pp. 1034–1041. IEEE (2009)
279. Sarlos, T.: Improved approximation algorithms for large matrices via random projections. In: 47th Annual IEEE Symposium on Foundations of Computer Science (FOCS'06), pp. 143–152. IEEE (2006)
280. Scherer, K.: Psychological models of emotion. In: Borod J (ed.) The Neuropsychology of Emotion, pp. 137–162. Oxford University Press, New York (2000)
281. Scherer, K., Shorr, A., Johnstone, T.: Appraisal Processes in Emotion: Theory, Methods, Research. Oxford University Press, Canary (2001)
282. Scherer, K.R.: Adding the affective dimension: a new look in speech analysis and synthesis. In: ICSLP, Philadelphia, pp. 1808–1811 (1996)
283. Schleicher, R., Sundaram, S., Seebode, J.: Assessing audio clips on affective and semantic level to improve general applicability. In: Fortschritte der Akustik – DAGA, Berlin (2010)
284. Sebe, N., Tian, Q., Loupias, E., Lew, M.S., Huang, T.S.: Evaluation of salient point techniques. In: International Conference on Image and Video Retrieval, pp. 367–377. Springer, London (2002)
285. Shan, C., Gong, S., McOwan, P.W.: Beyond facial expressions: learning human emotion from body gestures. In: BMVC, Warwick, pp. 1–10 (2007)
286. Simons, M., Tonhauser, J., Beaver, D., Roberts, C.: What projects and why. In: Proceedings of Semantics and Linguistic Theory (SALT), Vancouver, vol. 20, pp. 309–327 (2010)
287. Singh, P.: The open mind common sense project. KurzweilAI.net (2002)
288. Siorpaes, K., Hepp, M.: Ontogame: weaving the semantic web by online games. In: ESWC, Tenerife, pp. 751–766 (2008)
289. Smith, E., DeCoster, J.: Dual-process models in social and cognitive psychology: conceptual integration and links to underlying memory systems. Personal. Soc. Psychol. Rev. **4**(2), 108–131 (2000)
290. Smith, J., Chang, S.: An image and video search engine for the world-wide web. In: Symposium on Electronic Imaging: Science and Technology, San Jose (1997)
291. Snyder, B., Barzilay, R.: Multiple aspect ranking using the good grief algorithm. In: HLT/NAACL, Rochester (2007)

292. Socher, R., Huval, B., Manning, C.D., Ng, A.Y.: Semantic compositionality through recursive matrix-vector spaces. In: EMNLP, Jeju Island, pp. 1201–1211. Association for Computational Linguistics (2012)
293. Socher, R., Perelygin, A., Wu, J.Y., Chuang, J., Manning, C.D., Ng, A.Y., Potts, C.: Recursive deep models for semantic compositionality over a sentiment treebank. In: EMNLP, Seattle (2013)
294. Somasundaran, S., Wiebe, J., Ruppenhofer, J.: Discourse level opinion interpretation. In: COLING, Manchester, pp. 801–808 (2008)
295. Sowa, J.: Semantic networks. In: Shapiro S. (ed.) Encyclopedia of Artificial Intelligence. Wiley, New York (1987)
296. Speer, R.: Open Mind Commons: an inquisitive approach to learning common sense. In: Workshop on Common Sense and Interactive Applications, Honolulu (2007)
297. Speer, R., Havasi, C.: ConceptNet 5: a large semantic network for relational knowledge. In: Hovy, E., Johnson, M., Hirst, G. (eds.) Theory and Applications of Natural Language Processing, chap. 6. Springer, Berlin (2012)
298. Speer, R., Havasi, C., Lieberman, H.: Analogyspace: reducing the dimensionality of common sense knowledge. In: AAAI (2008)
299. Srinivasan, U., Pfeiffer, S., Nepal, S., Lee, M., Gu, L., Barrass, S.: A survey of MPEG-1 audio, video and semantic analysis techniques. Multimed. Tools Appl. **27**(1), 105–141 (2005)
300. Stevenson, R., Mikels, J., James, T.: Characterization of the affective norms for English words by discrete emotional categories. Behav. Res. Methods **39**, 1020–1024 (2007)
301. Stork, D.: The open mind initiative. IEEE Intell. Syst. **14**(3), 16–20 (1999)
302. Strapparava, C., Valitutti, A.: WordNet-Affect: An affective extension of WordNet. In: LREC, Lisbon, pp. 1083–1086 (2004)
303. Strapparava, C., Valitutti, A.: Wordnet affect: an affective extension of wordnet. In: LREC, Lisbon, vol. 4, pp. 1083–1086 (2004)
304. Tang, D., Wei, F., Qin, B., Liu, T., Zhou, M.: Coooolll: a deep learning system for twitter sentiment classification. In: Proceedings of the 8th International Workshop on Semantic Evaluation (SemEval 2014), pp. 208–212 (2014)
305. Tang, D., Wei, F., Yang, N., Zhou, M., Liu, T., Qin, B.: Learning sentiment-specific word embedding for twitter sentiment classification. In: Proceedings of the 52nd Annual Meeting of the Association for Computational Linguistics, vol. 1, pp. 1555–1565 (2014)
306. Thaler, S., Siorpaes, K., Simperl, E., Hofer, C.: A survey on games for knowledge acquisition. Tech. rep., Semantic Technology Institute (2011)
307. Torrance, G., Thomas, W., Sackett, D.: A utility maximisation model for evaluation of health care programs. Health Serv. Res. **7**, 118–133 (1972)
308. Tracy, J., Robins, R., Tangney, J.: The Self-Conscious Emotions: Theory and Research. The Guilford Press (2007)
309. Tropp, J.A.: Improved analysis of the subsampled randomized Hadamard transform. Adv. Adapt. Data Anal. **3**(01n02), 115–126 (2011)
310. Turney, P.: Thumbs up or thumbs down? Semantic orientation applied to unsupervised classification of reviews. In: ACL, Philadelphia, pp. 417–424 (2002)
311. Turney, P., Littman, M.: Measuring praise and criticism: inference of semantic orientation from association. ACM Trans. Inf. Syst. **21**(4), 315–346 (2003)
312. Turney, P.D.: Thumbs up or thumbs down?: semantic orientation applied to unsupervised classification of reviews. In: Proceedings of the 40th Annual Meeting on Association for Computational Linguistics, pp. 417–424. Association for Computational Linguistics (2002)
313. Tversky, A.: Features of similarity. Psychological Review 84(4), 327–352 (1977)
314. Ueki, N., Morishima, S., Yamada, H., Harashima, H.: Expression analysis/synthesis system based on emotion space constructed by multilayered neural network. Syst. Comput. Jpn. **25**(13), 95–107 (1994)
315. Urban, J., Jose, J.: EGO: A personalized multimedia management and retrieval tool. Int. J. Intell. Syst. **21**(7), 725–745 (2006)
316. Urban, J., Jose, J., Van Rijsbergen, C.: An adaptive approach towards content-based image retrieval. Multimed. Tools Appl. **31**, 1–28 (2006)

317. Velikovich, L., Goldensohn, S., Hannan, K., McDonald, R.: The viability of web-derived polarity lexicons. In: NAACL, Los Angeles, pp. 777–785 (2010)
318. Vesterinen, E.: Affective computing. In: Digital Media Research Seminar, Helsinki (2001)
319. Vicente, L.: On the syntax of adversative coordination. Nat. Lang. Linguist. Theory **28**(2), 381–415 (2010)
320. Vogl, T.P., Mangis, J., Rigler, A., Zink, W., Alkon, D.: Accelerating the convergence of the back-propagation method. Biolog. Cybern. **59**(4–5), 257–263 (1988)
321. Ware, J.: Scales for measuring general health perceptions. Health Serv. Res. **11**, 396–415 (1976)
322. Ware, J., Kosinski, M., Keller, S.: A 12-item short-form health survey: construction of scales and preliminary tests of reliability and validity. Med. Care **34**(3), 220–233 (1996)
323. Ware, J., Sherbourne, C.: The MOS 36-item short-form health survey (SF-36). Conceptual framework and item selection. Med. Care **30**, 473–83 (1992)
324. Wessel, I., Merckelbach, H.: The impact of anxiety on memory for details in spider phobics. Appl. Cognit. Psychol. **11**, 223–231 (1997)
325. Westen, D.: Implications of developments in cognitive neuroscience for psychoanalytic psychotherapy. Harv. Rev. Psychiatry **10**(6), 369–73 (2002)
326. Whissell, C.: The dictionary of affect in language. Emot. Theory, Res. Exp. **4**, 113–131 (1989)
327. Wiebe, J.: Learning subjective adjectives from corpora. In: AAAI/IAAI, pp. 735–740 (2000)
328. Wiebe, J., Wilson, T., Cardie, C.: Annotating expressions of opinions and emotions in language. Lang. Resour. Eval. **39**(2), 165–210 (2005)
329. Wilson, T., Wiebe, J., Hoffmann, P.: Recognizing contextual polarity in phrase-level sentiment analysis. In: HLT/EMNLP, Vancouver, pp. 347–354 (2005)
330. Wilson, T., Wiebe, J., Hwa, R.: Just how mad are you? Finding strong and weak opinion clauses. In: AAAI, San Jose, pp. 761–769 (2004)
331. Winston, P.: Learning structural descriptions from examples. In: Winston, P.H. (ed.) The Psychology of Computer Vision, pp. 157–209. McGraw-Hill, New York (1975)
332. Winterstein, G.: What but-sentences argue for: a modern argumentative analysis of but. Lingua **122**(15), 1864–1885 (2012)
333. Wu, H.H., Tsai, A.C.R., Tsai, R.T.H., Hsu, J.Y.J.: Sentiment value propagation for an integral sentiment dictionary based on commonsense knowledge. In: 2011 International Conference on Technologies and Applications of Artificial Intelligence (TAAI), Taoyuan, pp. 75–81. IEEE (2011)
334. Xia, R., Zong, C., Hu, X., Cambria, E.: Feature ensemble plus sample selection: domain adaptation for sentiment classification (extended abstract). In: IJCAI, Buenos Aires, pp. 4229–4233 (2015)
335. Xia, Y., Cambria, E., Hussain, A., Zhao, H.: Word polarity disambiguation using bayesian model and opinion-level features. Cognit. Comput. **7**(3), 369–380 (2015)
336. Yan, J., Yu, S.Y.: Magic bullet: a dual-purpose computer game. In: ACM SIGKDD, Paris, pp. 32–33 (2009)
337. Yang, C., Lin, K.H.Y., Chen, H.H.: Building emotion lexicon from weblog corpora. In: Proceedings of the 45th Annual Meeting of the ACL on Interactive Poster and Demonstration Sessions, Prague, pp. 133–136. Association for Computational Linguistics (2007)
338. Yu, H., Hatzivassiloglou, V.: Towards answering opinion questions: separating facts from opinions and identifying the polarity of opinion sentences. In: EMNLP, Sapporo, pp. 129–136. ACL (2003)
339. Zeki, S., Romaya, J.: Neural correlates of hate. PloS One **3**(10), 35–56 (2008)
340. Zellig, H.: Distributional structure. Word **10**, 146–162 (1954)
341. Zeng, Z., Tu, J., Liu, M., Huang, T.S., Pianfetti, B., Roth, D., Levinson, S.: Audio-visual affect recognition. IEEE Trans. Multimed. **9**(2), 424–428 (2007)
342. Zirn, C., Niepert, M., Stuckenschmidt, H., Strube, M.: Fine-grained sentiment analysis with structural features. In: IJCNLP, Chiang Mai (2011)
343. van Zwol, R., Garcia, L., Ramirez, G., Sigurbjornsson, B., Labad, M.: Video tag game. In: WWW, Beijing (2008)

Index

© Springer International Publishing Switzerland 2015

E. Cambria, A. Hussain, *Sentic Computing*, Socio-Affective Computing 1,
DOI 10.1007/978-3-319-23654-4

Printed by Printforce, the Netherlands